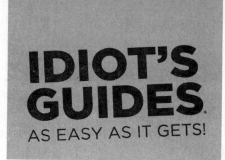

IDIOT'S GUIDES.
AS EASY AS IT GETS!

P9-DXI-610

Algebra I

by Carolyn Wheater

ALPHA
A member of Penguin Random House LLC

For my daughter.

ALPHA BOOKS

Published by Penguin Random House LLC

Penguin Random House LLC, 375 Hudson Street, New York, New York 10014, USA • Penguin Random House LLC (Canada), 90 Eglinton Avenue East, Suite 700, Toronto, Ontario M4P 2Y3, Canada (a division of Pearson Penguin Canada Inc.) • Penguin Books Ltd., 80 Strand, London WC2R 0RL, England • Penguin Ireland, 25 St. Stephen's Green, Dublin 2, Ireland (a division of Penguin Books Ltd.) • Penguin Random House LLC (Australia), 250 Camberwell Road, Camberwell, Victoria 3124, Australia (a division of Pearson Australia Group Pty. Ltd.) • Penguin Books India Pvt. Ltd., 11 Community Centre, Panchsheel Park, New Delhi—110 017, India • Penguin Random House LLC (NZ), 67 Apollo Drive, Rosedale, North Shore, Auckland 1311, New Zealand (a division of Pearson New Zealand Ltd.) • Penguin Books (South Africa) (Pty.) Ltd., 24 Sturdee Avenue, Rosebank, Johannesburg 2196, South Africa • Penguin Books Ltd., Registered Offices: 80 Strand, London WC2R 0RL, England

IDIOT'S GUIDES and Design are trademarks of Penguin Random House LLC

International Standard Book Number: 978-1-61564-775-0
Library of Congress Catalog Card Number: 2014957357

20 19 18   8 7 6 5 4 3 2

Interpretation of the printing code: The rightmost number of the first series of numbers is the year of the book's printing; the rightmost number of the second series of numbers is the number of the book's printing. For example, a printing code of 15-1 shows that the first printing occurred in 2015.

Printed in the United States of America

Note: This publication contains the opinions and ideas of its author. It is intended to provide helpful and informative material on the subject matter covered. It is sold with the understanding that the author and publisher are not engaged in rendering professional services in the book. If the reader requires personal assistance or advice, a competent professional should be consulted. The author and publisher specifically disclaim any responsibility for any liability, loss, or risk, personal or otherwise, which is incurred as a consequence, directly or indirectly, of the use and application of any of the contents of this book.

Most Alpha books are available at special quantity discounts for bulk purchases for sales promotions, premiums, fund-raising, or educational use. Special books, or book excerpts, can also be created to fit specific needs. For details, write: Special Markets, Alpha Books, 375 Hudson Street, New York, NY 10014.

Publisher: *Mike Sanders*
Associate Publisher: *Billy Fields*
Senior Acquisitions Editor: *Lori Cates Hand*
Development Editor: *Ann Barton*
Senior Production Editor: *Janette Lynn*

Cover Designer: *Laura Merriman*
Book Designer: *William Thomas*
Indexer: *Brad Herriman*
Layout: *Brian Massey*
Proofreader: *Laura Caddell*

Contents

Introduction

Do you have a favorite type of reading? Do you love romance or biography? What about movies and television? Is there a genre that you prefer? Are you a science fiction buff, or a reality show junkie? I enjoy a lot of those, but if I have to pick a favorite type of story to read or watch, I'll go for mysteries every time. Within the general heading of mysteries, my true passion is police procedurals, or in common language, cop shows. The ones that involve great forensic science are particular favorites but anything that involves investigation will get my attention.

You might be wondering why I'm talking about books and movies in an algebra book. You can probably find quite a few people who would describe algebra as a mystery, and of course, with all the problems that instruct us to "find x," algebra can feel like an endless search. The real connection, I think, is between the structure of the story and the structure of this book.

How This Book Is Organized

This book is presented in five parts. Each part contains from one to five chapters, grouped together around some common aspects of the work.

Part 1, Introduction to Algebra, takes you from arithmetic to the fundamentals of algebra. This part provides a foundation by looking at numbers, introducing variables, and showing how your knowledge of arithmetic can transfer to algebra. This part also introduces functions, the relationships at the center of much of our work in algebra. That's the set up to our mystery story. We get a little background and an introduction to our task.

Part 2, Linear Relationships, gets you started on the straight and narrow. Mysteries always begin with a smooth start to the investigation, and our study of algebra begins with linear functions that operate in clear, predictable ways. We graph lines, find the equations of lines from their graphs, and solve linear equations and inequalities. It's all connected and tied up with a few clear rules, but it's enough for us to solve quite a few problems.

Part 3, Variations on the Line, introduces the lines that don't behave like other lines, the horizontal and vertical lines. You'll meet absolute value functions, made up of parts of two different lines, and learn to solve systems, problems that simultaneously solve two linear equations.

Part 4, Polynomials, is that crucial part of the mystery story where everything starts to look complicated but the heroes untangle it all. When we first meet polynomials, they look complicated, but we'll bring our algebra skills to bear, and learn to add, subtract, multiply, and divide them, and even disassemble them into products of simpler expressions.

Part 5, Radical, Quadratic, and Rational Functions, gives you the tools to amaze all onlookers by using your previously acquired skills to unravel the mysteries of functions, which that seem dramatically different from the linear relationships with which we began.

Extras

As you make your way through this investigation of algebra, you'll see some items set off in ways meant to catch your attention. Here's a summary of what you'll see.

 DEFINITION

All the evidence in the world won't help you solve a mystery if you don't know what it means. Clear definitions are important to understanding. The Definition sidebars throughout this book will show you critical words and phrases that you'll want to know and use.

 ALGEBRA TRAP

Like most things, algebra is full of spots where it's easy to trip up. Don't get caught making a rookie mistake! These sidebars serve as a caution and try to help you think and act like an algebra detective.

 TIP

Algebra is full of little bits of information that are helpful to know, but many times we don't learn them until we've had some experience. Watch for sidebars throughout the book that point out those bits of information to give you the benefit of others' experience.

THINK ABOUT IT

These sidebars will encourage you to stop for a moment to think about why the techniques you're learning work. You'll find that when you understand the "whys," you'll have a better grasp of the concepts and of when and how to use them.

 CHECK POINT

As you investigate the various mysteries that algebra presents, you'll want to stop from time to time and take stock of what you know. Check Points ask you to answer a few questions to see if you're ready to move on. You'll find the answers for these Check Point questions in Appendix B.

But Wait! There's More!

Have you logged on to idiotsguides.com lately? If you haven't, go there now! As a bonus to the book, we've included 400 practice problems to help you practice your new algebra skills. You can find them at idiotsguides.com/algebra1.

Acknowledgments

It's always hard to know who to mention at this point in a book. You, my intended reader, may have no idea who these people are, and you may skip over this section because of that. Or you may read this and wonder if these folks are as strange as I am. The most important people may never see the book, and yet they should be mentioned.

So I'll begin by being forever grateful to Irma Jarcho, who welcomed me to my first teaching job by giving me a view of the task I was undertaking that struck fear into my heart. I stepped into my first classroom much better prepared because of Irma, and under her guidance, I grew into a teacher. She was my supervisor, we were peers, and toward the end, I was her supervisor, but through all the years, she was a friend I could count on for wisdom and unvarnished truth. Thank you, Irma, for teaching me what I was capable of doing and being. Rest in peace.

My gratitude goes to Grace Freedson, of Grace Freedson's Publishing Network, who not only won't let me get lazy, but also offers me projects, like this one, that are satisfying and challenging, and help me to grow as a teacher and as a person. My thanks also go to Lori Hand and Ann Barton, for making this project an absolute delight, from start to finish, and for making my scribblings about math look good and make sense.

One of the things I tell my students is that it's normal, natural, even valuable, to make mistakes. It's how we learn. More correctly, correcting our mistakes is how we learn. We all make mistakes. The Technical Reviewer's job is to read everything I've written about the math in this book and make sure it's correct and clear. That job also includes checking all the problems, and the answers, and finding my mistakes. Yes, I make mistakes, and I am grateful to my Technical Reviewer, Meredith Brown McNamara, for finding them and pointing them out to me. Her corrections and her advice make this a better book for you. Meredith and I were fellow teachers, and I was impressed with her dedication, vision, and creativity in redesigning our Geometry course. I count myself fortunate to know her and to work with her.

Personal thanks go to Pat Taranto, who keeps me grounded and balanced, and Elise Palazzi, who teaches me to relax. Finally, to my family who patiently put up with my obsessive approach to every new project, a giant thank you. To my daughter, Laura Wheater, to my siblings, Betty and Tom Connolly, and Frank and Elly Catapano, and all the nieces and nephews, thank you for teaching me to take a break, to celebrate, and to remember that life is more than work.

Trademarks

All terms mentioned in this book that are known to be or are suspected of being trademarks or service marks have been appropriately capitalized. Alpha Books and Penguin Random House LLC cannot attest to the accuracy of this information. Use of a term in this book should not be regarded as affecting the validity of any trademark or service mark.

Introduction to Algebra

All my favorite detective stories begin by setting the scene, letting us know where we are, and at least a little bit of how we got there. Once the crime has been committed or the mystery has been introduced, the principal characters begin to gather the essential information: who, what, where, when, how, and why. Of course, that's never enough, or it would be a very short story, but it is a good way to start our investigation of algebra.

In these first three chapters, we'll look at a little bit of background of arithmetic that gives rise to algebra, and build some basic understandings and skills that are the foundation for a study of algebra. We'll transfer the ideas of arithmetic to a world that includes variables, and explore the essential concept of a function, the relationship at the heart of much algebraic work.

Arithmetic to Algebra

Before a doctor can learn to diagnose an illness, she must understand the workings of the human body. Before a forensic scientist can analyze evidence, he must have a thorough background in the sciences. Anyone who wants to tackle the challenges of algebra needs first to understand the system of arithmetic from which algebra grows. So let's begin with arithmetic, and then move on to algebra. In this chapter, we'll look at some key elements of how arithmetic works: numbers, operations, rules, and shortcuts. That will be a good review and also give us a chance to talk about why it works the way it does.

In This Chapter

- Understanding the workings of our number system
- Reviewing the rules for signed numbers
- Practicing shortcuts for exponents
- Applying the order of operations

The Real Numbers

The set of numbers you use for most of your work is a set called the *real numbers*. This set is made up of several other sets of numbers.

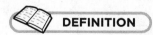 **DEFINITION**

The **rational numbers** are all numbers, positive or negative, that can be written as a fraction. Any number that doesn't fit that description is **irrational**. The **real numbers** are all numbers; positive and negative, rational and irrational.

The Real Numbers

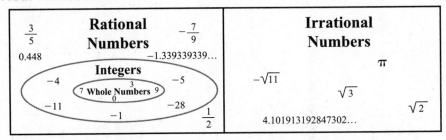

Early in your experience with arithmetic, you work with just whole numbers. Gradually, your mathematical world expands to include fractions and decimals, and negative as well as positive numbers. At that point, you're using the *rational numbers*.

The whole numbers, made up of zero and the counting numbers, form the set {0, 1, 2, 3, 4 …}. Include the opposites of those counting numbers, and you have the integers: {…-4, -3, -2, -1, 0, 1, 2, 3, 4…}. The rational numbers are made up of all numbers that can be written as a ratio, or fraction, of two integers, so they include all the positive and negative fractions, and all the positive and negative decimals that are equivalent to fractions.

 TIP

The … in a set of numbers means that the list continues forever. If there's a pattern to what you can see of the list, the pattern continues.

Eventually, however, you realize that there are decimals that aren't equivalent to fractions. When you change a fraction to a decimal by dividing the numerator by the denominator, one of two things happens: either the decimal terminates, as $\frac{1}{4} = 0.25$ terminates, or the decimal goes on forever, repeating a pattern, as $\frac{1}{3} = 0.3333…$ repeats. When you realize that there are decimals that go on forever but don't repeat, like 0.112123123412345…, you realize there are irrational numbers. The rational numbers and the irrational numbers together make the real numbers.

The real numbers are basically all the numbers you use, all the numbers you can think of, but surprisingly, not all the numbers you can imagine. There are other numbers mathematicians think about that don't fit our image of real numbers, and they are called the *imaginary numbers*. We'll take a little peek at them later on. For now, we're working with the real numbers.

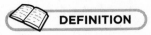 **DEFINITION**

Imaginary numbers are numbers that don't fit in the real number system, but that mathematicians use to solve certain types of problems.

Basic Operations

Most people would say that there are four basic operations of arithmetic: addition, subtraction, multiplication, and division. They'd be right, but other people would say there are only two basic operations, addition and multiplication, and they'd be right, too. We could go even further and say that multiplication is just repeated addition, so addition is the one basic operation. The point of this is not to take away operations, but to narrow our focus to what is essential. We'll take the middle ground and say that addition and multiplication are the basic operations.

Addition, the first operation most people learn, is a putting together, a shortcut for counting. You learn it with the whole numbers, but then you have to adapt the rules to deal with negative numbers and fractions and decimals. Once you've encountered the integers, however, subtraction can be defined as adding the opposite, a special case of addition. Multiplication is a shortcut for repeated addition, but so useful that we'll call it a separate operation, and again adjust the rules for integers and rational numbers. When your world of numbers includes fractions, division can be defined as multiplying by the reciprocal, and so a special case of multiplication.

Other Operations

There are other operations used in arithmetic, the principle ones being powers and roots. A *power* is a shortcut for repeated multiplication. The *base* is the number being multiplied, and the small raised number, called an *exponent*, tells how many times it appears as a factor. Taking a *root* is the opposite, or inverse, of raising to a power. If 2, raised to the third power, is 8, then the third root of 8 is 2.

 **DEFINITION**

A **power** is a way to tell how many times a number should be used in repeated multiplication. The number to be multiplied is the **base** of the power, and the small raised number that tells how many times to use it is called the **exponent**. A **root** is the opposite operation of a power, a way of undoing a power.

Roots are one of the places where you encounter irrational numbers, and where mathematicians began to think about those imaginary numbers. There are gaps between integers that are perfect squares. You know that $3^2 = 9$, so $\sqrt{9} = 3$, and $4^2 = 16$, so $\sqrt{16} = 4$, but what about all the numbers between 9 and 16? They have square roots, and those square roots are numbers between 3 and 4, but when you try to look for them, you realize they can't all be written as fractions. They're not rational numbers.

Properties

The operations of addition and multiplication work within certain rules or properties. Most of these properties talk about one operation on one set of numbers. We'll look at the real numbers and investigate each property.

The *closure property* says that all your work with the real numbers stays in the real numbers. Whenever you add two real numbers, you get a real number. If you multiply two real numbers, you get a real number. Every time you perform the operation on two members of the set, you get another member of the set. The real numbers are closed for addition and for multiplication.

 THINK ABOUT IT

The closure property doesn't seem to say much until you look at where it doesn't work. For example, the integers are not closed for division. Two divided by three is not an integer. The positive numbers are not closed under subtraction.

The *associative property* says that grouping can be changed without changing the result. The associative property for addition of real numbers tells you that $(3 + 8) + 4 = 3 + (8 + 4)$. The associative property for multiplication of real numbers says that the product $(2 \times 5) \times 7$ is the same as the product $2 \times (5 \times 7)$. In a series of the same operation, you can choose where to start. You don't have to start at the left and move to the right if you see an easier move.

The *commutative property* guarantees that when you add two or more real numbers, or multiply two or more real numbers, the result is the same, even if the order of the numbers is changed. The sum of $6 + 8$ is the same as $8 + 6$ and multiplying 9×4 gives the same answer as 4×9. In addition or multiplication, a change in the order of the numbers will not change the outcome.

It's important to remember that addition and multiplication are commutative but subtraction and division are not. $5 - 1$ is not the same as $1 - 5$ and $8 \div 2$ doesn't equal $2 \div 8$. However, you can think of subtraction as adding the opposite. For example, $5 - 1$ is $5 + -1$ and that is equal to $-1 + 5$. The same is true for division. Change a division like $8 \div 2$ to $8 \cdot \frac{1}{2}$ and that's equal to $\frac{1}{2} \cdot 8$.

The *identity property* assures that there is some number in the set that leaves other numbers unchanged when you operate with it. For the real numbers under addition, this number is 0. For the real numbers under multiplication, this number is 1. Adding 0 or multiplying by 1 changes nothing. Zero is the additive identity. One is the multiplicative identity.

Following on the heels of the identity property, the *inverse property* says there is a way to get back to the identity element. For every number in the set, except 0, there is an opposite, or additive inverse, you can add to get to the additive identity of 0. For every real number except 0, there is a reciprocal, or multiplicative inverse, that you can multiply by to produce an answer of 1, the multiplicative identity.

THINK ABOUT IT

You could say that zero is its own opposite because 0 + 0 = 0, but usually zero is considered a special case, neither positive nor negative. Zero has no reciprocal. There is no number that multiplies with zero to make 1, or for that matter, any number but zero.

There is one property that involves both addition and multiplication, and that's the *distributive property* for multiplication over addition. This property looks at adding two real numbers and then multiplying the result by a third real number, for example, 8(5 + 2). The property tells you that you can produce the same answer by multiplying each of the addends by the multiplier first, and then adding. In other words, 8(5 + 2) = 8(5) + 8(2).

ALGEBRA TRAP

Be aware that you can't distribute addition over multiplication. 4 + (5 × 3) does not equal (4 + 5) × (4 + 3).

Closure, identity, and inverse are not properties that are talked about a lot, but they guarantee that the system of arithmetic operates logically. They make it possible to undo arithmetic that's been performed, and that will be crucial for algebra. The associative, commutative, and distributive properties pop up regularly in day-to-day arithmetic. You can use them to rearrange a problem to make the arithmetic simpler, and you'll continue to do that in algebra.

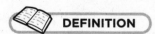

DEFINITION

The properties of the number system include:

Closure: Adding or multiplying two numbers of the set gives us another number in the set.

Associative: $(a+b)+c = a+(b+c)$, $(ab)c = a(bc)$

Commutative: $a+b = b+a$, $ab = ba$

Identity: Adding 0 or multiplying by 1 leaves a number unchanged.

Inverse: Every number, except 0, has an opposite and a reciprocal. Adding the opposite brings you back to 0. Multiplying by the reciprocal gets back to 1. Zero is its own opposite but has no reciprocal.

Distributive: $c(a+b) = ca + cb$

CHECK POINT

Choose all the correct answers for each question.

1. $-\dfrac{5}{13}$ is
   a) an integer
   b) a rational number
   c) an irrational number
   d) a real number

2. $\sqrt{49}$ is
   a) an integer
   b) a rational number
   c) an irrational number
   d) a real number

3. $\sqrt{5}$ is
   a) an integer
   b) a rational number
   c) an irrational number
   d) a real number

4. $\sqrt{6.25}$ is
   a) an integer
   b) a rational number
   c) an irrational number
   d) a real number

Name the property illustrated by each statement.

5. $-3(4 + -4) = (-3 \cdot 4) + (-3 \cdot -4)$

6. $-12 + 12 = 0$

7. $93 + 26 + 17 + 14 = 93 + 17 + 26 + 14$

8. 1,274 times 1,939 is a real number.

9. $17 \cdot 1 = 17$

10. $\left(\dfrac{2}{5} \cdot \dfrac{3}{25}\right) \cdot \dfrac{125}{6} = \dfrac{2}{5} \cdot \left(\dfrac{3}{25} \cdot \dfrac{125}{6}\right)$

Integers and Absolute Value

Working with integers is not terribly different from the arithmetic of whole numbers; in fact, the arithmetic of fractions is much more to learn. What you've learned about operations with integers carries over to all real numbers, so reviewing that information is worthwhile.

What Are the Integers?

Integers are positive and negative whole numbers and zero. If you imagine the whole numbers placed along a line to the right of zero, evenly spaced, and then hold a mirror at zero, the reflection in the mirror will give you a picture of the negative numbers.

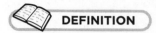

> The **integers** are the whole numbers and their opposites. All the spaces between the integers are filled with rational and irrational numbers.

This "through the looking glass" quality of the negative integers can make it confusing when you're trying to talk about size, order, and comparison. Remembering that numbers to the left are smaller will help with order and size. When you're trying to compare two numbers, the one on the left on the number line is smaller. For arithmetic operations, separating the absolute value of a number from its sign can sometimes be helpful.

Absolute Value

The *absolute value* of a number is the value of the number without its sign. The number 12 and the number -12 have the same absolute value. Both are 12 units away from zero, but 12 is to the right of zero and -12 to the left.

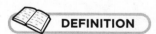

> The **absolute value** of a number is its distance from zero without regard to direction.

Addition (and Subtraction)

Addition and subtraction can be understood as moving right or left along the number line, but once that understanding is in place, you'll want some shortcuts so that you don't have to build a giant number line to do your taxes. Separating the absolute value of a number from its sign helps to express the simple version of those shortcuts.

To add two numbers with the same sign, add the absolute values and keep the sign. A positive number plus a positive number is a positive number, and a negative plus a negative is a negative.

To add two numbers with different signs, subtract the absolute values and take the sign of the number with the larger absolute value. You have competing forces, one pulling right and one pulling left. They'll partially cancel each other out, but the one with the larger absolute value will win.

When you add 12 + -7, the absolute values are 12 and 7, and you subtract them to get 5. The 12 is the bigger absolute value, so the sign of the 12 is the sign of the answer: 12 + -7 = 5. When you add 4 + -9, the absolute values are 4 and 9, and the answer will be negative, because the larger absolute value is the 9, and that started out negative. 4 + -9 = -5.

If subtracting means adding the opposite (for example, to subtract 9 you actually add -9), then subtracting signed numbers becomes a special case of the addition rule. To subtract an integer from another number, add the opposite of the integer and follow the rules for addition.

Multiplication (and Division)

Multiplication is repeated addition. If you add 7 + 7 + 7, you're adding three 7s, and that's the multiplication problem 3 × 7 = 21. Applying that logic to adding several negative numbers, like -4 + -4 + -4 + -4 + -4, you have 5 × -4 and you can see that it gives you -20.

You can use the idea of repeated addition to see that the product of two positive numbers is a positive number, and that the product of a positive and a negative is negative. What happens when you multiply a negative number by a negative number? One way to think of such a problem is to see each of the negative numbers as -1 times a positive number; for example, $-6 \cdot -8$ is $(-1 \cdot 6) \cdot (-1 \cdot 8)$. Then you can use the associative and commutative properties to rearrange.

$$-6 \cdot -8 = (-1 \cdot 6) \cdot (-1 \cdot 8)$$
$$= -1 \cdot (6 \cdot -1) \cdot 8$$
$$= -1 \cdot (-1 \cdot 6) \cdot 8$$
$$= (-1 \cdot -1) \cdot (6 \cdot 8)$$
$$= 1 \cdot 48$$
$$= 48$$

THINK ABOUT IT

Some of the properties of real numbers can also help explain why the product of two negative numbers must be positive. You know that adding a number and its opposite will give you 0, and you know that 0 multiplied by any number, positive or negative, is 0. Let's say your number is 5. $5 + -5 = 0$. Now multiply by a negative number like -3. By the distributive property, $-3(5 + -5) = -3 \cdot 5 + -3 \cdot -5 = -15 + -3 \cdot -5$, but the final answer must be 0, so the $-3 \cdot -5$ must be positive 15.

The result of multiplying a negative number by a negative number is a positive number. Two negatives, like two positives, give a positive. That lets us formulate two rules.

1. To multiply two numbers with the same sign, multiply the absolute values, and make the product positive.

2. To multiply two numbers with different signs, multiply the absolute values and make the product negative.

Since division is defined as multiplying by the reciprocal, the same sign rules apply. To divide by an integer, multiply by its reciprocal, following the rules for multiplication.

CHECK POINT

Simplify each expression. Use the associative and commutative properties whenever they're helpful.

11. $4 + -17 + -3 + 29$

12. $-7 \cdot 5 \cdot -8 \cdot 12$

13. $8 \cdot -6 \cdot 3 \cdot -4$

14. $-14 + 8 + -31 + 27 + 6 + -4$

15. $-25(18 + -14) + 8(-9 + 12)$

16. $7(-39 + 43) + 14 \cdot -2$

These expressions involve real numbers that are not all integers, but the same rules apply.

17. $\dfrac{13}{5} + -\dfrac{17}{3} + -\dfrac{3}{5} + \dfrac{29}{3}$

18. $4 \cdot (-0.6) \cdot (1.2) \cdot (-10)^2$

19. $-1.4 + 8.3 + -3.1 + 2.7 + -4.5$

20. $-14\left(-\dfrac{3}{8} + \dfrac{3}{7}\right) + \left(-\dfrac{7}{4} + \dfrac{7}{2}\right)$

Properties of Exponents

The basic operations of addition and multiplication are themselves shortcuts. Addition is a shortcut for counting, and multiplication is a shortcut for repeated addition. One of the other common operations, raising a number to a power, is a shortcut for repeated multiplication.

Meaning of Exponent

The product $2 \cdot 2 \cdot 2$ can be represented using an exponent as 2^3. The 2, the number actually being used as a multiplier or factor, is called the *base*. The small raised number, called an *exponent*, tells you how many 2s to multiply. The base and the exponent together form a power. The power 5^4 is shorthand for multiplying four 5s. The power 8^2 tells you to use 8 as a factor twice.

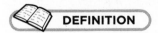 **DEFINITION**

The official name for the operation of raising to a power is **exponentiation**.

Powers show up frequently in algebra, so it's useful to have some shortcuts for dealing with them. There are three essential rules that will make it easier to operate with powers, but it's important to state, right up front, that these rules only apply to powers with the same base. If you need to multiply 2^3 by 5^2, you'll just have to work out that it's 8 times 25 and end up with 200. There's no shortcut when the bases are different. Notice, too, that there's no shortcut for addition or subtraction of powers. You just need to work those out.

 TIP

When you raise a negative number to a power, be sure to put the negative number in parentheses before putting the exponent on. This is especially important when using a calculator. If you mean $(-3) \cdot (-3)$, write $(-3)^2$. If you write -3^2, your reader (or your calculator) will see that as $-(3 \cdot 3)$.

Multiplying Powers

When you multiply powers of the same base, you're joining two statements of repeated addition. If you multiply $7^3 \cdot 7^5$, you're saying $(7 \cdot 7 \cdot 7) \cdot (7 \cdot 7 \cdot 7 \cdot 7 \cdot 7)$. That's all multiplication, so you can wrap the whole job into one exponent. You want to multiply a total of eight 7s, so you can write 7^8. Because the bases are the same, you just condense two sets of 7s, three in one and five in the other, into one set of eight.

To multiply powers of the same base, keep the base and add the exponents.

Dividing Powers

To understand the rule for dividing powers, you need to think about fractions for a moment. If you write a quotient of two powers of the same base, say $\frac{3^4}{3^2}$, you have a fraction whose numerator and denominator have factors in common: $\frac{3^4}{3^2} = \frac{3 \cdot 3 \cdot 3 \cdot 3}{3 \cdot 3}$. In that situation, you have the right to simplify the fraction by removing factors equivalent to 1. $\frac{3^4}{3^2} = \frac{3 \cdot 3 \cdot 3 \cdot 3}{3 \cdot 3} = \frac{\cancel{3}}{\cancel{3}} \cdot \frac{\cancel{3}}{\cancel{3}} \cdot \frac{3 \cdot 3}{1} = 3^2$. Each 3 in the denominator cancels a 3 in the numerator, leaving $4 - 2$, or two 3s.

To divide powers of the same base, keep the base and subtract the exponents. Subtract the exponent of the denominator from the exponent of the numerator.

Raising a Power to a Power

Raising a power to a power is repeated multiplication of repeated multiplication, and that interpretation will give us the shortcut. To raise 2^5 to the third power means to use 2^5 as a factor three times. So $\left(2^5\right)^3 = \left(2^5\right) \cdot \left(2^5\right) \cdot \left(2^5\right)$. You could write out exactly what that means and then count up how many 2s you have, or you could use the rule for multiplying powers of the same base. $\left(2^5\right)^3 = \left(2^5\right) \cdot \left(2^5\right) \cdot \left(2^5\right) = 2^{5+5+5} = 2^{15}$. (That's 32,768, by the way.)

To raise a power to a power, keep the base and multiply the exponents.

Zero Exponents and Negative Exponents

Because an exponent tells you how many times to use the base as a factor, most of the exponents you encounter are positive integers. But applying the rules for exponents can lead to some results that might be puzzling.

When you apply the rule for dividing powers to a problem like $\frac{6^3}{6^3}$, you find yourself with $\frac{6^3}{6^3} = 6^{3-3} = 6^0$. To understand what it means to use a factor zero times, you go back to the problem and apply some basic arithmetic. When you divide $\frac{6^3}{6^3}$, you're dividing a number $6^3 = 216$ by itself, and a number divided by itself equals one. That gives us a definition of a zero exponent.

Any non-zero number to the zero power is one.

THINK ABOUT IT

Why do we say "any non-zero number"? The definition of a zero exponent relies on division, and division by zero is impossible.

The division rule can produce another unexpected result. If you divide $\frac{4^2}{4^3}$, the division rule says $\frac{4^2}{4^3} = 4^{2-3} = 4^{-1}$. To make sense of that, go back to basic arithmetic. $\frac{4^2}{4^3} = \frac{4 \cdot 4}{4 \cdot 4 \cdot 4} = \frac{\cancel{4}}{\cancel{4}} \cdot \frac{\cancel{4}}{\cancel{4}} \cdot \frac{1}{4} = \frac{1}{4}$.

Is it a coincidence that you wind up with 1 over the base of the power? Not at all. There are more 4s in the denominator than the numerator; in this case, the denominator has one extra 4.

A base to the -1 power is equivalent to the fraction with a numerator of 1 and a denominator equal to the base.

If you have more than one extra factor in the denominator, for example, $\frac{5^3}{5^7}$, instead of getting your base to the -1 power, you'll get some other negative power, in this case, $\frac{5^3}{5^7} = 5^{-4}$. Look again at the arithmetic to see what that means. $\frac{5^3}{5^7} = \frac{5 \cdot 5 \cdot 5}{5 \cdot 5 \cdot 5 \cdot 5 \cdot 5 \cdot 5 \cdot 5} = \frac{\cancel{5}}{\cancel{5}} \cdot \frac{\cancel{5}}{\cancel{5}} \cdot \frac{\cancel{5}}{\cancel{5}} \cdot \frac{1}{5 \cdot 5 \cdot 5 \cdot 5} = \frac{1}{5^4}$. An exponent of -4 indicates the fourth power of the base will be the denominator, with a numerator of 1.

A negative exponent on a non-zero base represents a fraction with a numerator of 1 and a denominator of the base raised to the corresponding positive power.

Fractional Exponents

You might be wondering, now that we've looked at zero and negative exponents, if there can be fractional exponents. The answer is yes, and the short explanation is that fractional exponents represent roots.

Imagine you have a power, like 8^3. Taking the third root of 8^3 would undo the third power and get you back to 8. If there is an exponent that represents the third root, it should be true that raising 8^3 to that power equals 8, or $\left(8^3\right)^? = 8^1$. The power of a power rule says you should multiply those exponents, so 3 times what number will give you an exponent of 1? The missing number is $\frac{1}{3}$, so an exponent of $\frac{1}{3}$ represents the third root.

The third root of 8^3 would be written as $\sqrt[3]{8^3}$. The little 3 in the crook of the radical, the index, says you're undoing the third power. (If there's no index, it's a square root.) The exponent version would be $\left(8^3\right)^{\frac{1}{3}}$. Both the radical version and the exponent form are equal to 8. $\sqrt[3]{3^8} = \left(3^3\right)^{\frac{1}{3}} = 8$.

Any exponent of the form $\dfrac{1}{\text{a number}}$ represents a root. The number in the denominator of the fraction tells you what root.

CHECK POINT

Use the rules for exponents to simplify each expression.

21. $\left(2^3\right) \cdot \left(2^2\right)$

22. $\dfrac{5^4}{5^2}$

23. $\left(3^2\right)^3$

24. 12^{-1}

25. $\dfrac{4^5}{4^7}$

26. $6^{-3} \cdot 6^5$

27. $\dfrac{81^{14}}{81^{14}}$

28. $\left(5^7\right)\left(5^{-8}\right)$

29. $\dfrac{9^4}{3 \cdot 9^3}$

30. $\left(\dfrac{7^5 \cdot 11^{13}}{7^{12} \cdot 11^9}\right)^0$

Until now, we've been primarily focusing on one operation at a time: addition, or multiplication, or exponentiation. However, you don't usually encounter operations one at a time. You're often asked to add and multiply, or find powers and subtract, or do all of those in the same problem. How do you keep it organized?

Order of Operations

Faced with the problem $16 + 5 \cdot 12$, it's important that everyone agrees on the order in which the arithmetic should be done, or we could get different answers to the same problem. Working left to right might seem logical, and would give you $16 + 5 \cdot 12 = 21 \cdot 12 = 252$, but if you thought that the multiplication was easier, and did that first, you'd get $16 + 5 \cdot 12 = 16 + 60 = 76$. They can't both be right, so which one is? The answer is an agreement that multiplication will come before addition. (The correct answer to the problem above is 76.)

The agreed-upon order of operations states that multiplication and division are done first, from left to right, followed by addition and subtraction, from left to right. However, if there are grouping symbols, such as parentheses, those come before anything else. If the problem had been written (16 + 15) × 12, you would do the operation in parentheses first.

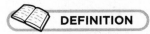 **DEFINITION**

The **order of operations** is an agreement about the sequence in which arithmetic will be performed. Expressions in parentheses are simplified first, then powers. Multiplication and division are done as met from left to right, and finally addition and subtraction, from left to right.

The associative and commutative properties allow us to rearrange an addition or a multiplication to make the arithmetic easier. When addition and multiplication appear in the same problem, there are often places to apply the distributive property. The computation $5(16+24)+8(92+18)$ can be done by following the signal from the parentheses and adding first.

$$5(16+24)+8(92+18) = 5(40)+8(110) = 200+880 = 1,080$$

But if you find it easier, you can distribute the multiplication.

$$5(16+24)+8(92+18) = 5 \cdot 16 + 5 \cdot 24 + 8 \cdot 92 + 8 \cdot 18$$
$$= 80 + 120 + 736 + 144$$
$$= 1,080$$

You also have the option to mix your strategies. $5(16+24)$ is not too hard to calculate by either method, but I know what $5 \cdot 16$ and $5 \cdot 24$ equal, so I'd probably distribute there. On the other hand, $8(92+18)$ is much easier if I add first.

Just as multiplication, a shortcut for repeated addition, gets done before addition, exponents, the symbol for repeated multiplication, need to be handled before multiplication or addition. Do exponents, multiplication, and addition, in that order, unless parentheses say otherwise. What about subtraction and division? Those are defined as special cases of addition and multiplication, respectively, so subtraction gets done with addition and multiplication with division. That leaves us with a rule for the order of operations that you can remember as PEMDAS: parentheses, exponents, multiplication (and division), addition (and subtraction).

CHECK POINT

Simplify each expression, following the order of operations.

31. $4-3\cdot4+3$

32. $(4-3)\cdot(4+3)$

33. $4-3\cdot(4+3)$

34. $(4-3)\cdot4+3$

35. $5^2-2\cdot(7-3)$

36. $(66-54)\div3+10\div5-(6-2^2)$

37. $5-2+2\cdot(8-5)\div2$

38. $17-(8-3)\cdot(2+1)\div5$

39. $8\left(2(4+3)^2-20+12\right)\div4$

40. $7+3^2\div2-6$

A clear understanding of arithmetic is the foundation you need to begin your study of algebra. Working with unknowns will require you to think carefully not only about what kind of arithmetic you need to do but also about how that arithmetic works and why.

The Least You Need to Know

- Arithmetic in the real number system is governed by the associative, commutative, distributive, identity, inverse, and closure properties.
- The sign rules for operating on integers hold true for all real numbers: rational and irrational.
- Shortcuts for multiplying, dividing, and finding powers of powers simplify work with exponents
- The order of operations—parentheses, exponents, multiplication and division, addition and subtraction—governs arithmetic, and is represented by the acronym PEMDAS.

Variables and Expressions

What makes a detective story interesting is that some important piece of the story is unknown. There wouldn't be much need for a detective if we already had all the facts. That element of the unknown is also what distinguishes algebra from arithmetic.

In arithmetic, you have all the numbers and you know what operations need to be performed; you just have to do the math. When there's a missing piece—a number you don't know or one that keeps changing—you need algebra to solve that mystery. In this chapter, we'll bring variables into arithmetic, and investigate what changes and what stays the same when one (or more) of the numbers you're working with is unknown.

In This Chapter

- Understanding and using variables
- Identifying and combining like terms
- Simplifying expressions with the distributive property
- Multiplying variable expressions

Variables and Expressions

A *variable* is any symbol used to stand for an unknown number or for a number that keeps changing. It can be as simple as a blank to fill in, as in $4 + \underline{} = 7$, but in algebra, variables usually take the form of letters. If you order a new jacket that costs $45 but have to pay for shipping, you could use a variable like x to stand for the shipping charge, and the total cost of your purchase would be $(45 + x)$. If you find out what the shipping charge is, you can do the math and know exactly what the total will be, but until you have that information, a variable lets you talk about it even though one piece is unknown.

 **DEFINITION**

A **variable** is a letter or other symbol that takes the place of an unknown number.

Because a variable takes the place of a number, you can write phrases and sentences using variables as you would numbers. Making the transition from words to symbols like numbers, variables, and operations can take some practice. It's a lot like translating from one language to another. When you first try it, you translate word by word, but as you become more comfortable with the language, you're able to read the whole phrase or sentence and say the same thing in the new language.

If you write, in words, "six plus nine equals fifteen," you can translate that to the symbols $6 + 9 = 15$. If you write the words "six plus some number equals fifteen," you can use a variable, like n, to represent "some number" and the sentence becomes $6 + n = 15$. The translation is word by word. Six becomes 6, plus becomes +, and so on.

Whenever you're trying to translate, it's helpful to understand the grammar of the language. In algebra, numbers and variables are nouns and operation signs like + and × are conjunctions that tie the numbers and variables together. In English, you learn to distinguish between a phrase, such as "the weather forecast for Monday" and a sentence, such as "The weather forecast for Monday is sunny." The key difference between them is the verb. In algebra, equal signs and inequality signs are the verbs. If you write $6 + n$, that's a phrase made up of two nouns and a conjunction: "six and a number." In algebra, that's called an expression. If you write $6 + n = 15$, the equal sign acts as a verb. Your sentence is "six plus some number is 15." Equations and inequalities are the sentences of algebra.

Language provides many different ways to say the same thing, and employing that variety can make writing more interesting. The symbolic representations used in math are less diverse, however, and as a result many different sentences translate to the same set of symbols. The sentence "six plus nine equals fifteen" could also be "the sum of six and nine is fifteen" or "nine more than six is fifteen" or "six increased by nine is fifteen." All those sentences would translate to $6 + 9 = 15$.

It's difficult, if not impossible, to list all the different ways to express the same mathematical statement, but here are some common ones.

Operation	Common Expressions	Example
Addition	-4 added to some number y	$-4+y$
	6 increased by a number x	$6+x$
	2 more than a number n	$2+n$
	The sum of a number t and 9	$t+9$
Subtraction	A number y decreased by 5	$y-5$
	The difference of a number z and 11	$z-11$
	19 less than a number x	$x-19$
	A number y reduced by 4	$y-4$
Multiplication	The product of a number n and 7	$n \cdot 7$
	Eleven times a number y	$11y$
	-3 times a number x	$-3x$
Division	The quotient of z and 6	$\dfrac{z}{6}$
	12 divided by a number x	$\dfrac{12}{x}$
	The ratio of a number t and 10	$\dfrac{t}{10}$

Because addition and multiplication are commutative, the order of the numbers can be changed. $x+4$ and $4+x$ are equivalent. For subtraction and division, which are not commutative, take the order of the numbers from the sentence. The quotient of a and b is $\dfrac{a}{b}$. The quotient of b and a is $\dfrac{b}{a}$. The difference of x and y is $x-y$ but the difference of y and x is $y-x$.

 ALGEBRA TRAP

There is one exception to the rule for subtraction. When it says something like "9 less than a number," you subtract 9 from the number. If you say it with numbers rather than variables, it's easier to see. A number that's two less than five is clearly three, or $5-2$, not $2-5$. If you want a number that's 2 less than x, it's $x-2$, not $2-x$.

Of course, the context of the question can change the way the information is presented to you. If you're looking for two numbers that add to 100 and one is greater than the other by 25, you could call one number x and the other $x + 25$. Then your sentence is $x + x + 25 = 100$. If you paid for two lamps plus a $25 delivery fee, and your total bill was $100, that would also be $x + x + 25 = 100$. Different situations (in words) can lead to the same set of symbols.

CHECK POINT

Write an equation for each sentence, using the specified variable.

1. The product of -4 and a number x.

2. A number y increased by 18.

3. The quotient of a number z and 8.

4. The product of 9 and a number x, decreased by 3.

5. Nineteen more than the product of -3 and a number x.

6. The difference of two numbers, a and b, multiplied by -5.

7. A number x is subtracted from 19.

8. Twice a number x increased by 4 is 8 more than a number y.

9. George bought 3 hot dogs and spent $1.50 for a drink. His total bill was $12. (Use h for the price of one hot dog.)

10. One number is 15 more than another and their sum is 42. (Use n for the smaller number.)

To talk intelligently about algebra, we need to be speaking the same language, using the same words to mean the same things. We've defined a variable as a symbol, usually a letter, which takes the place of a number. Let's build from there.

Factors, Terms, and Expressions

When you multiply, each number or variable is a *factor* and the result is a *product*. When you multiply two numbers, you can find the product and give the result as a single number. When one or more of the factors are variables, you can't arrive at a single number because you don't know the value of the variable, but you can express the multiplication in a compact fashion.

Multiplying the number 7 and the variable x can be written as $7x$. Multiplying x times x, or any variable by itself, can be condensed by using an exponent. $x \cdot x = x^2$. Multiplying two different variables, like x and y, just becomes xy.

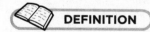 **DEFINITION**

A **product** is the result of multiplying. Each of the numbers or variables multiplied is a **factor**.

A collection of factors, whether numbers or variables or both, that are connected only by multiplication is called a *term*. A term may contain a power because exponents are a shortcut for multiplication, but not addition. When you add two or more terms, you create an *expression*. Expression is a more general word. Terms are expressions, and sums, differences, and quotients of terms are also expressions.

In writing, you build sentences from phrases, and the phrases are made up of words. In algebra, the sentences are equations or inequalities. The equal signs and inequality signs act like verbs. Expressions are phrases and terms are like words.

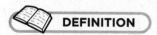 **DEFINITION**

An **expression** is any mathematical calculation. An expression containing numbers and variables combined only by multiplication is called a **term**.

The expression $-3x$ is a term with two factors: -3 and x. Technically, each of these factors is a *coefficient*, the -3 is the numerical coefficient and the x is the variable coefficient. If you look at the roots of the word coefficient, it means things that work together. The -3 and the x work together to make the term.

The expression $5x^2$ is a term. It has a coefficient of 5. The variable is squared, which means the term is really $5 \cdot x \cdot x$, so technically it has three factors.

The expression $5x^2 + -3x$ (or $5x^2 - 3x$) is the sum of two terms.

The expression $\dfrac{5x^2 - 3x}{x^2 + 2x + 4}$ is a quotient of two expressions. The numerator has two terms and the denominator has three terms.

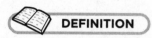 **DEFINITION**

The word **coefficient** usually refers to the numerical part of a term. Technically, a term is made up of a numerical coefficient and a variable coefficient.

CHECK POINT

Identify the number of factors in each term.

11. $3x$

12. $-6y^2$

13. $23xy$

14. $-48x^3y^2$

Identify the number of terms in each expression.

15. $x^2 + 3x - 8$

16. $t - 2$

17. $6x^2$

18. $9x^2 - 4x$

19. $\dfrac{5x}{7-x}$

20. $\dfrac{x^2 - 2x}{x^2 + 6x - 9}$

Arithmetic with Variables

When you add, each of the numbers or variables being added is called an *addend* and the result is a *sum*. At first, it might seem that arithmetic with variables would be impossible. How can you add numbers you don't know? Although you can't assign a number to $x + x$, you can say that it's two of the same number, and that means $x + x = 2x$. You still can't say what it's worth until you know what x stands for, but you have a simpler way to represent the addition.

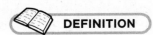

DEFINITION

Each of the numbers or variables in an addition problem is called an **addend**, and the result is called a **sum**.

You add variable terms that are *like*, terms that have the same variable and same exponent, by just adding their coefficients. When you write a term like $5x$, 5 times x means five copies of the number you don't know. It means $x + x + x + x + x$. If you add $3x$ to that, you're adding another three copies of that unknown number, $x + x + x$. That will give you a total of eight of those unknown numbers, eight x's or $8x$.

DEFINITION

Like terms have the same variable and the same exponent.

As long as the addends are just copies of the same variable, all you're really doing is counting up how many you have. When you start to add terms, like $6x$ and $-8x$, you'll want a simpler method than just counting. As long as the terms are like, that is, terms with the same variable and same exponent, you can just add the coefficients and keep the variable parts the same.

In addition to combining terms with the exact same variable part, you can also combine the numbers that have no variable part. These terms, which are just numbers, are called *constant terms* or simply constants.

 **DEFINITION**

Constants are simple terms containing only numbers and no variables.

Combining Like Terms

You'll want to simplify expressions whenever you can. You could write the expression $5x-1+3x+5-2x+8+9x-4$, which is the sum of eight terms. That expression can be simplified, however. The simplified result will be much easier to write and definitely easier to work with. You can use the associative and commutative properties to rearrange, and then do as much of the arithmetic as you can without knowing what number x stands for. The rule to remember is that you can only combine terms that have identical variable parts.

$$5x-1+3x+5-2x+8+9x-4$$
$$\left(5x+3x-2x+9x\right)+\left(-1+5+8-4\right)$$
$$15x+8$$

Simplifying an expression as complex as $5x-1+3x+5-2x+8+9x-4$ requires using the associative and commutative properties to group like terms together. You wouldn't usually show every single step in that rearrangement, but it's worthwhile to think about them once or twice. Working left to right, your first step would be $5x-1$ but those are unlike, so use the associative property to regroup. $\left(5x-1\right)+3x+5-2x+8+9x-4=5x+\left(-1+3x\right)+5-2x+8+9x-4$. Use the commutative property to change the order. $5x+\left(3x-1\right)+5-2x+8+9x-4$. Regroup again. $\left(5x+3x\right)-1+5-2x+8+9x-4$. Now you can combine the like variable terms, and the constants that follow. The more terms, the more regrouping and switching required.

Remember to think of subtraction as adding the opposite. When you see $5x-1$, you want to think of that as $5x+-1$. When you think of subtraction as adding the opposite, it lets you use the associative and commutative properties to rearrange the expression for convenient simplifying.

CHECK POINT

Simplify each expression by combining like terms.

21. $(x+3)+(9+x)$

22. $(2x-1)+(3x+9)$

23. $(x^2+5)+(x^2+6x)+(8x+3)$

24. $(2x+7)+(8x+4)+(-x^2-1)$

25. $(2x-4)+(6x+1)+(3-x)$

26. $(y+5)+(2y-9)+(4y+10)+(y-6)$

27. $(8t-4)+(-t-1)+(13t-7)+(-2t-9)$

28. $(x^2+4)+(-x^2-1)+(-3-5x)+(10x+9)$

29. $(3x^2+2x+4)+(x^2-3x+10)$

30. $(4x^2-5x-6)+(9+7x-4x^2)$

When you multiply, each number or variable is a factor and the result is a product. Multiplication with variable expressions can have different looks: one term times one term, one term times an expression with more than one term, and multiplication of two expressions that both have more than one term. All the different cases break down to multiplying one term by one term, and doing that means multiplying the coefficients, multiplying the variables, and applying the rules for exponents when necessary. For example, $4x \cdot -2x^2 = (4 \cdot -2) \cdot (x \cdot x^2) = -8x^3$.

The Distributive Property

When we talked about the basic properties of arithmetic, we said that the distributive property allows you to choose to add and then multiply, or multiply and then add. You can do $6(5+3)$ either as $6(5+3) = 6(8) = 48$ or as $6(5+3) = 6(5)+6(3) = 30+18 = 48$. It's up to you to choose.

In algebra, when you're dealing with unknowns, the distributive property gives you a chance to simplify expressions that present an obstacle. When you face $-4(x+7)$, you can't add first. The x and the 7 are unlike terms and can't be combined, and you don't know the value of x. But thanks to the distributive property, you can multiply first.

$$-4(x+7) = (-4)x + (-4)(7) = -4x - 28$$

When you use the distributive property with a multiplier that's a constant, you just need to be careful about rules for signs. When you use the distributive property with a multiplier that includes a variable, you'll also need to remember the rules for exponents, primarily the rule for multiplying powers. To multiply $2x\left(x^2+3x+5\right)$, first distribute the $2x$.

$$2x\left(x^2+3x+5\right)=2x\left(x^2\right)+2x\left(3x\right)+2x\left(5\right)$$

Then do each individual multiplication, remembering that $x=x^1$ and that when you multiply powers of the same base, you keep the base and add the exponents.

$$\begin{aligned}2x\left(x^2+3x+5\right)&=2x\left(x^2\right)+2x\left(3x\right)+2x\left(5\right)\\&=\left(2x^1\cdot x^2\right)+\left(2\cdot3\cdot x^1\cdot x^1\right)+\left(2\cdot5\cdot x\right)\\&=2x^3+6x^2+10x\end{aligned}$$

 ALGEBRA TRAP

Remember that terms with the same variable but different exponents are unlike terms. An x is not the same as an x^2. If x stands for 7, x^2 stands for 49, definitely not the same thing. Like terms must have the exact same variable part, including the exponent.

When a negative number or a term with a negative coefficient needs to be distributed over an expression, it's important to pay attention to the signs. In the expression $-3x\left(x^2-5x+2\right)$, you need to multiply each of the three terms in the parentheses by $-3x$, and that means you'll need to be thinking carefully about the sign of each resulting term. $-3x\cdot x^2$ will be negative, specifically, $-3x^3$. $-3x\cdot-5x$ will be positive and $-3x\cdot2$ will be negative. You should get $-3x\left(x^2-5x+2\right)=-3x^3+15x^2-6x$.

When you see just a negative sign in front of an expression in parentheses, you're actually seeing an instruction to distribute -1. The effect is to change all the signs.

$$\begin{aligned}-\left(x^2-3x-4\right)&=-1\left(x^2-3x-4\right)\\&=\left(-1\cdot x^2\right)+\left(-1\cdot-3x\right)+\left(-1\cdot-4\right)\\&=-x^2+3x+4\end{aligned}$$

Notice that the x^2 became $-x^2$, and the $-3x$ and the -4 both became positive.

CHECK POINT

Simplify each expression by applying the distributive property (where needed) and combining like terms.

31. $2(x-4)+7(x+4)$

32. $-3(y+8)+8(y-3)$

33. $6(2t+5)+9(t-2)$

34. $-7(3w-5)+4(w+5)$

35. $x(x+4)-(x+4)$

36. $-8x(x-3)+6(x-4)$

37. $5a(2a-7)-2(8-a)$

38. $-4y(7-3y)-4(3-7y)$

39. $3x(2x+3y)+3y(2x+3y)$

40. $2x^2(x^2-7x+9)+3x(x^2-7x+9)+2(x^2-7x+9)$

The distributive property will let you simplify a multiplication of one term times an expression with multiple terms, but what if you need to multiply two larger expressions? What if you need to multiply an expression with two terms by an expression with three terms?

What would you do in arithmetic if you needed to multiply 429 by 32? You'd probably arrange the numbers one under the other, and start by multiplying 429 by 2, working right to left. Then you'd put a zero on the next line to shift things over one place, and multiply 429 by 3, again right to left. Finally you'd add the two lines. This should look familiar.

$$
\begin{array}{r}
429 \\
\times 32 \\
\hline
858 \\
12{,}870 \\
\hline
13{,}728
\end{array}
$$

You can do something similar with variable expressions. To multiply $4x^2 + 2x + 9$ by $3x + 2$, you can:

Arrange the expressions one under another. Put the one with more terms on top.

$$\begin{array}{r} 4x^2 + 2x + 9 \\ \underline{3x + 2} \end{array}$$

Multiply $4x^2 + 2x + 9$ by 2. Don't worry about carrying. Just multiply each term by 2.

$$\begin{array}{r} 4x^2 + 2x + 9 \\ \underline{3x + 2} \\ 8x^2 + 4x + 18 \end{array}$$

Put a zero under the 18.

$$\begin{array}{r} 4x^2 + 2x + 9 \\ \underline{3x + 2} \\ 8x^2 + 4x + 18 \\ \underline{0} \end{array}$$

Multiply $4x^2 + 2x + 9$ by $3x$. Remember the rules for exponents. Put the result on the same line as the zero.

$$4x^2 + 2x + 9$$

Add the two lines. Like terms should already be lined up. Here's how it should look.

$$\begin{array}{r} 4x^2 + 2x + 9 \\ \underline{3x + 2} \\ 8x^2 + 4x + 18 \\ \underline{12x^3 + 6x^2 + 27x + 0} \\ 12x^3 + 14x^2 + 31x + 18 \end{array}$$

This vertical multiplication takes time and careful attention, but it follows the same pattern you used in arithmetic. You can use it to multiply any two sums of terms. As is often the case, the most common problems have a shortcut.

The FOIL Rule

You will frequently find that you need to multiply the sum of two terms by another expression that is the sum of two terms. If you do this by the vertical method, you will see that this takes a total of four little multiplications. Each term of the first expression gets multiplied by each term of the second expression.

$$
\begin{array}{r}
3x + 5 \\
2x - 9 \\
\hline
-27x - 45 \\
6x^2 + 10x + 0 \\
\hline
6x^2 - 17x - 45
\end{array}
$$

The $3x$ times the $2x$ created the $6x^2$.

The $3x$ times the -9 created the $-27x$.

The 5 times the $2x$ created the $10x$.

The 5 times the -9 created the -45.

These four multiplications can be labeled using the acronym FOIL: First, Outer, Inner, Last. Start by writing the problem on one line: $(3x + 5)(2x - 9)$.

The $3x$ and the $2x$ are the FIRST terms in each expression.

The $3x$ and the -9 are the OUTER ends of the problem.

The 5 and the $2x$ are the INNER terms.

The 5 and the -9 are the LAST terms in each expression.

Do these four multiplications and combine any like terms.

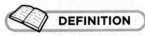 **DEFINITION**

The acronym **FOIL** stands for First, Outer, Inner, Last. It summarizes the four multiplications you need to do when multiplying two terms by two terms.

To multiply $(6x+5)(x-3)$ think of it as $(6x+5)(x+-3)$ and follow the FOIL rule.

First: $6x \cdot x = 6x^2$

Outer: $6x \cdot -3 = -18x$

Inner: $5 \cdot x = 5x$

Last: $5(-3) = -15$

So $(6x+5)(x-3) = 6x^2 - 18x + 5x - 15 = 6x^2 - 13x - 15$, once you combine the like terms $-18x$ and $5x$ in the middle.

CHECK POINT

Multiply using the vertical multiplication method.

41. $(2x-7) \cdot (x+4)$

42. $(x^2 + x + 2) \cdot (x-3)$

43. $(x^2 + 7x + 3) \cdot (2x+1)$

44. $(2x^2 - 3x - 8) \cdot (3x-4)$

Multiply using the FOIL rule.

45. $(x+3) \cdot (x+2)$

46. $(x+4) \cdot (x-2)$

47. $(2x+1) \cdot (x+7)$

48. $(4x-1) \cdot (2x+3)$

49. $(5x-2) \cdot (2x-3)$

50. $(x^2 + x) \cdot (x-3)$

The Least You Need to Know

- You can translate phrases about unknown numbers into expressions using variables.
- Terms contain numbers and variables connected only by multiplication.
- Like terms are terms that have the same variable(s) raised to the same power.
- You can add and subtract only like terms.
- Multiply terms using the rules for exponents.
- Use the distributive property to multiply one term by an expression, the FOIL rule to multiply two terms by two terms, and set up larger expressions in the vertical arrangement.

Functions

If you walk for an hour every day, what distance do you cover? Was your first thought "it depends"? It probably was, and rightfully so. The distance you cover in an hour depends upon the speed at which you walk, and whether you walk at a consistent speed for the entire hour. If you walk at a steady three miles per hour today, you'll cover three miles in your hour-long walk. If tomorrow you step up your pace, you might cover four miles in the same time. The distance varies and depends on your speed. In this chapter, we'll look at a way of representing relationships like this one with the help of variables.

Although most of the world thinks of variables as things that are unknown and need to be found, the name "variable" comes from the word *vary*. Variables are symbols used to represent quantities or values that vary, or are different at different times.

In This Chapter

- What is a function?
- How to quickly identify functions
- Using function notation
- Constructing tables and graphs to represent functions

What Is a Function?

Before we can talk about functions, we need to define another word that has a mathematical meaning a bit different from its use in ordinary speech. In English, if you talk about a relation, you're probably talking about a member of your family, someone who is related to you. In algebra, a *relation* is a pairing of the numbers from two sets.

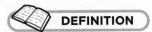 **DEFINITION**

A **relation** is a pairing of numbers from one set, called the domain, with numbers from another set, called the range.

If set A contains the numbers $\{1, 2, 3, 4, 5\}$ and set B contains $\{9, 12, 15, 18, 21\}$, you could create a relation by pairing each number from set A with a number from set B. You could make the pairs $\{(1, 9), (2, 12), (3, 15), (4, 18), (5, 21)\}$, but there's no rule that you have to go in order. You could pair them up as $\{(1, 15), (2, 9), (3, 12), (4, 21), (5, 18)\}$. Any matching up of numbers from set A with numbers from set B is a relation. The first set, set A in this example, is called the domain and the second set, set B, is called the range.

If a relation is a way of relating numbers from different sets, what is a function? It sounds like it's something that has a job to do, and often it does, but the definition of function is simpler than that. A *function* is a relation in which each element of the domain has only one partner from the range.

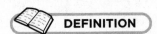 **DEFINITION**

A **function** is a relation in which each number in the domain has only one partner from the range.

Take a minute to think about that. Most people, reading that sentence, think all relations are functions, but they're not. If the only requirement I have is to pair the numbers, I could use the sets above and do something like this: $\{(1, 9), (1, 15), (2, 12), (2, 9), (3, 15), (3, 12), (4, 18), (4, 21), (5, 21), (5, 18)\}$.

I've paired them up, but I've used numbers from the domain twice, so each member of the domain has two different partners. That's a relation, but it's not a function. If even one number from the domain gets two different partners, it's not a function.

You might be thinking, "Okay, fine, but when does that ever really happen?" Suppose I take a set of numbers like $\{2,5,-3,1.2,-2,4,9\}$ and I pair each of those numbers with its square. I get $\{(2,4),(5,25),(-3,9),(1.2,1.44),(-2,4),(4,16),(9,81)\}$.

That's a function. Each number has one partner. But suppose I switch that relationship around. I take a set of numbers like $\{1,4,9,16\}$ and I pair each input with a number I could square to get my input. You might at first think that my pairs would be $\{(1,1),(4,2),(9,3),(16,4)\}$ but I'd be within my rights to say the pairs are $\{(1,1),(1,-1),(4,2),(4,-2),(9,3),(9,-3),(16,4),(16,-4)\}$. Each of the numbers in my domain could be paired with a negative number as well as a positive number. Pairing them with both of their square roots would not be a function.

I used rules to pair up the numbers in those examples, both for the function and the relation. Many times there is a rule that explains the pairing in a relation or a function, but not every time. Sometimes there is a rule, but it's not obvious from the pairing. If I were to pair the ages of my family members with their social security numbers, it might look like random numbers to you even though I followed a clear rule.

Because functions often have rules, you can imagine them as machines that have a particular job to do. The numbers in the domain are called *inputs*, and the numbers in the range are called the *outputs*. Imagine the inputs going into the function machine, the machine doing its job, and then generating an output for the range.

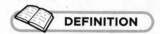

DEFINITION

The numbers from the domain that the function works on are **inputs**. The numbers in the range that the function produces are **outputs**.

CHECK POINT

Tell whether each set of pairs forms a function. Is there a rule that determines the pairs?

1.

Input	-2	-1	0	1	2
Output	-1	-2	2	5	3

2.

Input	-7	-5	3	4	8
Output	25	19	-5	-8	-20

3.

Input	-2	-1	0	1	2
Output	3	0	-1	0	3

4.

Input	-3	-1	5	7	5
Output	-4	-1	1	1	-2

5.

Input	4	9	16	25	36
Output	2	3	4	5	6

For each function, give the domain and range of the function. Does the function follow a rule?

6.

Input	-2	-1	0	1	2
Output	3	4	5	6	7

7.

Input	-2	-1	0	1	2
Output	4	4	4	4	4

8.

Input	0	2	4	6	8
Output	-3	1	5	9	13

9.

Input	10	20	30	40	50
Output	1	11	21	31	41

10.

Input	-5	6	8	11	21
Output	-4	-1	-7	12	-5

Function Notation

Functions can be specified by listing the pairs or by making a chart that shows the pairing. For large sets of numbers, this can become unwieldy, so there are other ways of communicating functions. This is where variables come in to play.

The numbers in the domain, the inputs, change. The input takes different values, so a variable can be used to stand for the input. This variable is often x, but it could be any variable. Once you choose a variable for the input, you try to express the rule for the function, if there is one, as an expression using that variable.

Suppose you want to write a simple function that matches each input with a number two units larger. You would need to say that you're going to match each number, x, with a number of the form $x + 2$. You could assign a variable to stand for the outputs, perhaps y, and say the output is two more than the input, or $y = x + 2$. When you input 5, the function will output 7. When $x = -2$, $y = 0$.

There's another way of writing this, called function notation, that has a couple of advantages. First, it gives the function a name. A letter, like f or g, is used to stand for the function. The variable that will stand for the inputs appears in parentheses, right after the function's name. The notation $f(x)$, which is read as "f of x," says that this is a function, its name is f and the variable x stands for its inputs. Then you give the rule, if there is one, so you might write $f(x) = x + 2$.

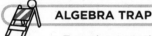

ALGEBRA TRAP

Function notation puts the variable that stands for inputs in parentheses after the name of the function. Don't confuse this with parentheses used to mean multiplication. $f(x)$ is not f times x. It's a function f that has inputs represented by x.

Naming the function this way makes it easier to talk and write about them. If I'm working with two functions, one that pairs inputs with two more than the input, and one that pairs inputs with their squares, having to repeat all that description every time I want to tell you about them can get tedious. If I write $f(x) = x + 2$ and $g(x) = x^2$, I can just talk about f and g, and you'll know which one I mean.

The other advantage to function notation is that it provides a compact way to say "find the output of this particular function for this particular input." Evaluating a function means finding the output for a particular input, and function notation lets you say "evaluate this" simply. If I want you to evaluate the function f above when the input is $x = 5$, I can just ask you to find $f(5)$. By writing the f, I've told you the name of the function you should use, and by replacing the x with 5, I've told you to make 5 the input. $f(5) = 5 + 2 = 7$. If I tell you to find $g(3)$, you know I want you to find out what the output will be when the input to the squaring function is 3. $g(3) = 3^2 = 9$.

CHECK POINT

Given the function $f(x) = 4 - 3x$, evaluate each of the following.

11. $f(3)$ 14. $f(-1)$

12. $f(-5)$ 15. $f(-100)$

13. $f(0)$

Evaluate each function as indicated.

16. If $g(x) = 8x - 2$ find $g(-4)$

17. If $h(x) = x^2 + 5$ find $h(0)$

18. If $p(x) = x^2 - 3x + 7$ find $p(-1)$

19. If $v(t) = 25 - 8t$ find $v(5)$

20. If $a(t) = 9.8$ find $a(20)$

Tables of Values

A function with a small domain can be defined by making a table showing all the pairs, as we did in the first Check Point of this chapter. If the domain of a function is large, making a table is impractical, and the function is generally defined by a rule. When a relation or a function defined by a rule needs to be evaluated for several different inputs, however, the inputs and outputs are often organized into a table of values.

 TIP

A function must send each input to only one output, so you shouldn't see repeated inputs. It's okay for outputs to repeat, however. You may see more than one input sent to the same output. There is even a type of function called a constant function in which every input goes to the same output.

In addition to giving you a simple way to organize the pairs, a table of values also provides a quick way of determining whether the relation you're looking at is a function. Just scan the inputs. If any input occurs more than once (and the outputs are different) the relation is not a function. If all the inputs are different, it's a function.

This relation is a function because each input is unique.

Input	-3	-1	0	1	3
Output	-7	1	5	9	17

This relation is not a function because you can see that the input -1 appears twice, once with an output of 1 and again with an output of -1.

Input	-3	-1	0	1	3	-1
Output	-7	1	5	9	17	-1

Having the inputs and outputs of a function arranged in a table of values can also help you find a rule for the function, if there is one. Make sure the inputs are in order, and if they're not, carefully rearrange the table. Be sure to keep each output with its input. Now look at the changes in the inputs and in the outputs. The following table shows the pairs of inputs and outputs for this function, arranged with inputs in order from smallest to largest.

ALGEBRA TRAP

When you're looking for a rule, you may tend to focus on outputs, and try to create rules that tell how to get from one output to the next. That may describe the pattern, but remember that the rule you want is how to get from an input to its output.

Change in input		+2		+1		+1		+2	
Input	-3	→	-1	→	0	→	1	→	3
Output	-7	→	1	→	5	→	9	→	17
Change in output		+8		+4		+4		+8	

Divide each change in output by the corresponding change in input. Do all the divisions come out the same? In this example, $\frac{8}{2} = \frac{4}{1} = \frac{4}{1} = \frac{8}{2}$. If all those divisions come out the same, as they do here, you're halfway to the rule. Take the result of the division, in this case 4, and multiply each input by that number.

Input	-3	-1	0	1	3
Input × 4	-12	-4	0	4	12
Output	-7	1	5	9	17
Output − (Input × 4)	$-7-(-12)$ $=-7+12$ $=5$	$1-(-4)$ $=1+4$ $=5$	$5-0$ $=5$	$9-4$ $=5$	$17-12$ $=5$

The inputs multiplied by four don't match the outputs. In this case, they're lower than the outputs by exactly five. The rule for this function is $f(x) = 4x + 5$, multiply the input by 4 and then add 5.

Here's a table of values for another function.

x	-3	-2	-1	0	1	2	3
$g(x)$	11	1	-5	-7	-5	1	11

When you look at the differences in the inputs and the outputs for this one, the ratios don't all match.

Change in input		+1		+1		+1		+1		+1		+1	
x	-3	$\rightarrow$	-2	$\rightarrow$	-1	$\rightarrow$	0	$\rightarrow$	1	$\rightarrow$	2	$\rightarrow$	3
$g(x)$	11	$\rightarrow$	1	$\rightarrow$	-5	$\rightarrow$	-7	$\rightarrow$	-5	$\rightarrow$	1	$\rightarrow$	11
Change in output		-10		-6		-2		+2		+6		+10	

Sometimes the changes don't match because there is no rule, but don't give up just yet. Do you notice the change in direction of the outputs? The outputs were going down, and then they went up. That change of direction is a sign that this function might have a rule that involves squaring.

Take another step by finding the changes in the changes in the outputs, like this.

Change in input		+1		+1		+1		+1		+1		+1	
x	-3	$\rightarrow$	-2	$\rightarrow$	-1	$\rightarrow$	0	$\rightarrow$	1	$\rightarrow$	2	$\rightarrow$	3
$g(x)$	11	$\rightarrow$	1	$\rightarrow$	-5	$\rightarrow$	-7	$\rightarrow$	-5	$\rightarrow$	1	$\rightarrow$	11
Change in output		-10		-6		-2		+2		+6		+10	
Change in change			+4		+4		+4		+4		+4		

If that second set of changes are all the same, your rule will start with half that number times x^2, in this case $2x^2$, and then you'll need to look for what to add or subtract.

x	-3	-2	-1	0	1	2	3
$2x^2$	18	8	2	0	2	8	18
$g(x)$	11	1	-5	-7	-5	1	11
Output $- 2x^2$	$11-18$ $=-7$	$1-8$ $=-7$	$-5-2$ $=-7$	$-7-0$ $=-7$	$-5-2$ $=-7$	$1-8$ $=-7$	$11-18$ $=-7$

In this case, the rule will be $g(x) = 2x^2 - 7$. This tactic won't find every possible rule, but it will help you find many of them. Sometimes just guessing and testing will help you find a rule. Sometimes you may think there's a rule because there seems to be a pattern, but you just don't have the skills yet to figure out what it is. But many times, you'll be able to express the rule for the function by following these steps.

CHECK POINT

Create a table of values for each function, using the given values for the inputs.

21. $f(x) = 3x - 1$ for values of x in the set $\{-2, -1, 0, 1, 2\}$.

22. $g(x) = 4 - x^2$ for values of x in the set $\{-5, -3, -1, 1, 3, 5\}$.

23. $p(z) = 2z + 9$ for values of z in the set $\{-4, -1, 0, 2, 5\}$.

24. $v(t) = t^2 - 1$ for values of t in the set $\{0, 1, 2, 3, 4, 5\}$.

25. $a(t) = -3t$ for values of t in the set $\{-5, -4, -3, -2, -1, 0\}$.

Each table of values below represents a function. Write a rule for the function, if possible, using function notation.

26.

t	0	1	2	3	4
$v(t)$	-3	2	7	12	17

27.

x	-1	0	1	2	3
$f(x)$	10	7	4	1	-2

28.

x	-3	-2	-1	0	1
$g(x)$	-4	1	4	5	4

29.

t	0	2	4	6	8
$a(t)$	2	5	8	11	14

30.

x	-4	-1	2	5	8
$p(x)$	-2	7	16	25	34

Graphs of Functions

Another common way of showing a function (or a relation) is to draw a picture of the function by plotting each pair of numbers as a point in the coordinate plane. The function in the table below could be shown as five points on the plane.

t	0	1	2	3	4
$v(t)$	-3	2	7	12	17

The five pairs, $\{(0,-3),(1,2),(2,7),(3,12),(4,17)\}$ can be plotted as five points, with their inputs counted on the horizontal axis and their outputs on the vertical axis. The graph looks like this.

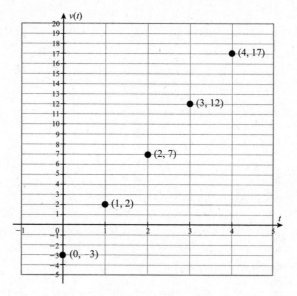

 TIP

The values from the domain are always on the horizontal axis, and the values from the range are on the vertical axis.

One advantage of presenting a function as a graph is that you can often spot a pattern faster in a graph. One look at the graph above and you can see that there's some sort of pattern or rule for this function. The points form a line. If there were no rule, the points would be just a random scattering of points.

The function below does not have a rule.

Input	-3	-1	5	7	8
Output	-4	-1	1	1	-2

When you plot the points $\{(-3,-4),(-1,-1),(5,1),(7,1),(8,-2)\}$ on a graph, you can quickly see that it's unlikely there would be one rule that explains these pairings.

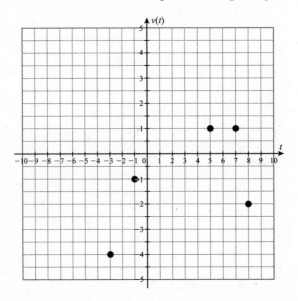

When there's a clear pattern to the graph, it's natural to want to connect the dots, to make a line or a curve that passes through all the pairs and highlights the pattern of the function. Sometimes it's appropriate to do that, but other times it's not.

Drawing that line says the function includes not only the pairs of numbers you plotted to see the pattern, but also all the pairs of numbers that name points on the line. That may or may not be true. If a function has a domain of all real numbers, and is defined by a rule, it's common to evaluate the function for a few values and plot those points, until you can see the pattern, and then connect the dots. But not every function has the whole set of real numbers as its domain.

THINK ABOUT IT

Division by zero is impossible, so if a variable appears in a denominator, you need to figure out what values of the variable might make that denominator equal zero, and eliminate them from the domain. It's also impossible to find the square root of a negative number in the real numbers, so if your function involves a square root, you need to eliminate any values of the variable that would give you a negative underneath the square root sign.

When you're wondering whether to connect the dots or not, you should read the information you're given carefully to see if you're told what the domain is. If it's all real numbers, you can feel free to connect. If the domain isn't stated, and you're given a table of values, assume that the function is just those pairs. Plot points, but don't connect. If the domain isn't stated, and you're given a rule, you can assume the domain is all numbers that can sensibly be substituted for the variable. Generally, that will be all real numbers, unless there's a variable in a denominator or under a square root sign.

 CHECK POINT

Draw a graph of each function. Should the graph be separate points or a continuous line or curve?

31.

x	0	-1	-4	-9	-16
$f(x)$	0	1	2	3	4

32. $f(x) = 2x - 3$

33. $g(x) = x^2 - 5$

34. $v(t) = 10 - t^2$

35. $a(t) = 5 - 3t$

36.

x	-2	-1	0	1	2
$f(x)$	2	1	0	1	2

37. $v(t) = -2t^2$

38. $f(x) = 5 - 4x$

39.

x	-3	-2	-1	0	1
$f(x)$	5	5	5	5	5

40. $g(x) = -2x$

Vertical Line Test

Presenting a relation as a graph also gives us a quick way to decide if the relation is a function, or more accurately to decide if it's not a function. A relation is not a function if you can find any input that has more than one output. When you're looking at a graph, that shows up as two (or more) points stacked vertically. The quick test is called the vertical line test.

If any vertical line crosses two or more points of the graph of a relation, the relation is not a function. Some people call this the pencil test, because you can take your pencil, hold it vertically and move it across the graph. If it ever crosses two points, the graph does not represent a function. The graph on the left is not a function. You can see several places were points are stacked in a vertical line. The graph on the right is a function.

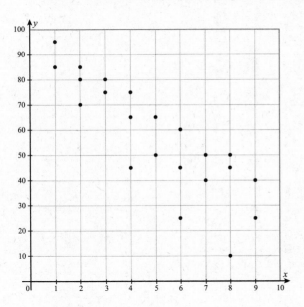 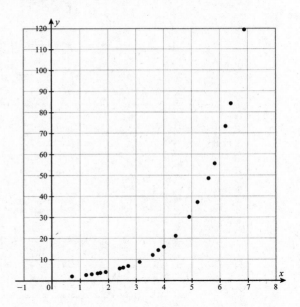

CHECK POINT

Use a vertical line test to determine whether each graph represents a function.

41.

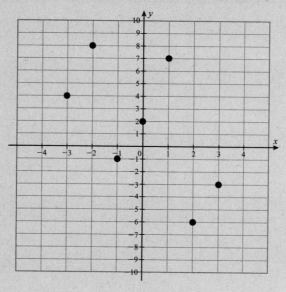

42.

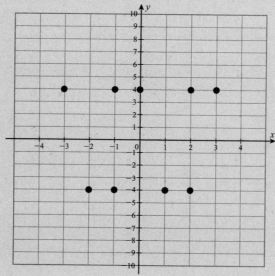

43.

44.

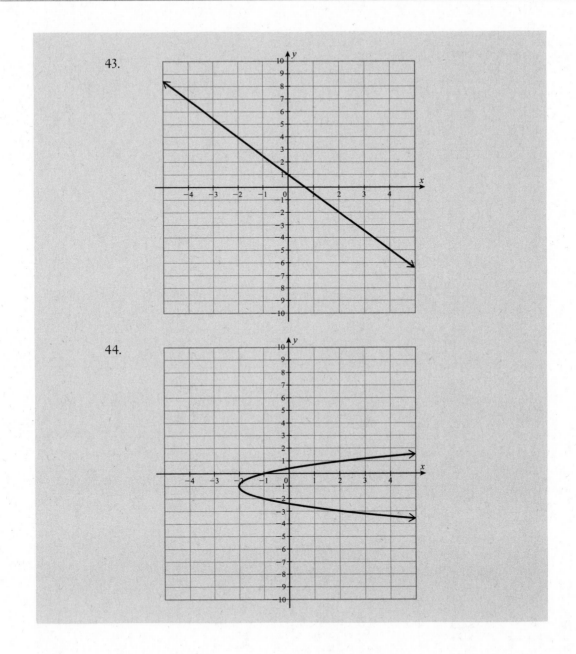

45.

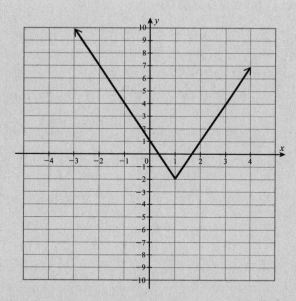

Construct a graph for each relationship below and use the vertical line test to determine if the relationship is a function.

46.

x	3	5	7	7	9
y	2	1	8	3	6

47.

x	-2	-1	0	1	2
y	4	1	0	1	4

48. $y = 3x^2 - 1$

49. $2x + 3y = 12$

50. $9x + y^2 = 4$

The Least You Need to Know

- A function is a pairing of numbers in which each input has only one output.

- The function notation $f(x)$ tells you that you have a function, that its name is f, and that the variable x stands for its inputs.

- You can show a function by a list of pairs, a table of values, a graph, or a rule that tells you how to calculate outputs.

- A graph that shows a function will pass the vertical line test: no vertical line crosses it twice.

- $f(3)$ means find the output of function f when the input is 3.

Linear Relationships

Have you noticed that at the beginning of a detective story, the investigation seems to move along just fine? The detectives gather their evidence and seem to be on a straight simple path to closing the case. That won't last, of course, but it always seems to start that way.

Algebra starts with a straight path, too. The beginning of algebra is a straight line study. We investigate functions whose graphs are lines, understand what makes an equation a linear function, and learn to solve linear equations and inequalities. That's the plan for Part 2 of this book, so get in line, and let's solve this case.

Equations

A great writer can craft a beautiful phrase, a combination of words that is so true, so perfect, so pleasing, that it stops you in your tracks. But if all that author ever wrote were phrases, there wouldn't be much of a story. No matter how beautiful the phrases are, eventually you have to put them together into sentences in order to communicate your thoughts.

Expressions, no matter how well they describe the mathematics you see happening, are just phrases. To truly communicate, you have to assemble expressions into mathematical sentences. In this chapter, we'll look at the most common mathematical sentence, the equation, and see how we can use equations to help us find the value of the unknowns the variables represent.

The phrase "the product of -4 and a number x" becomes the expression $-4x$. The sentence "The product of -4 and a number x is -8" becomes the equation $-4x = -8$. The equal sign serves as the verb in the mathematical sentence. The expression is on the left side of the equation and the number -8 is on the right side.

In This Chapter

- Understanding equations
- Simplifying to make solving easier
- Solving one- and two-step equations
- Using equations to describe situations

TIP

The relationship of equality is symmetric, which means that if $a = b$, then $b = a$. What that means for you is that the equations $-4x = -8$ and $-8 = -4x$ say exactly the same thing. Rewrite in whichever form you prefer.

Like expressions, equations vary from very simple to very complicated, but if you translate each phrase as you read it, you'll get the right equation. The sentence "The product of 9 and a number x, decreased by 3, is 10 times the number" becomes $9x - 3$ on the left side, the part before the "is" or equal sign, and $10x$ on the right. The equation is $9x - 3 = 10x$. Translating "The difference of a and 3, multiplied by -5, is zero" requires parentheses. It becomes $-5(a - 3) = 0$.

Simplifying Before Solving

When you approach an equation, you're eager to get to the unknown, to find out what number that variable is standing for. But it wouldn't make sense to try to get to the value of the variable without first stripping away some of the complications. In the case of equations, those complications take the form of expressions that can be simplified.

Before you begin to solve an equation, you want to simplify both sides of the equal sign as much as possible. Don't even think of it as an equation yet. Just see it as two expressions, treat each side as a separate job, and do any simplifying you can. For equations without exponents or roots, your goal should be to have no more than two terms on either side: a variable term and a constant term.

The equation $2(x - 7) + 5 = 3x + 1 + 7x$ isn't ready to solve. Focus on the left side, $2(x - 7) + 5$. Distribute the 2, and $2(x - 7) + 5$ becomes $2x - 14 + 5$. Then combine the like terms -14 and +5, and the left side of the equation becomes $2x - 9$. That's two terms, one variable term, $2x$, and a constant term, -9, so that side is now simplified enough. Turn your attention to the right side, $3x + 1 + 7x$. Combine the like terms, $3x$ and $7x$, and the right side becomes $10x + 1$. That's two terms, a variable term and a constant term. The equation that started as $2(x - 7) + 5 = 3x + 1 + 7x$ has turned into $2x - 9 = 10x + 1$. Each side has two terms, a variable term and a constant term. Now the equation is ready for solving. When you simplify, remember PEMDAS, the order of operations, and be sure to combine only like terms.

 ALGEBRA TRAP

When you simplify, remember the order of operations. $6 - 4(x + 1)$ is not $2(x + 1)$. The multiplication must be done before the subtraction. $6 - 4(x + 1) = 6 - 4x - 4 = 6 - 4 - 4x = 2 - 4x$.

CHECK POINT

Simplify the expression on each side of the equation. Leave no more than two terms on any side.

1. $x + (x+1) + (x+2) = 132$

2. $(2y-5) + (y+8) = 6 + (y-3)$

3. $4t - 2(t+5) = 8 + t - 3$

4. $2(x-4) = 3x - (x+1)$

5. $16 - (3a-7) = (8a-4) - 2(a+5)$

6. $x + (x+3) = 5(x+2) - 7$

7. $4(v-3) = 6 - 2(v+1)$

8. $5(x-3) - (x+2) = (6x-1) + 3(x+8)$

9. $-3(t+5) + 9t = 5(t+1) - (20-t)$

10. $4a - 7(a+3) + 2 = 6 + 3(1-a) - 5$

Solving One-Step Equations

The most complicated equation you should try to solve is one that has a variable term and a constant term on each side. Anything more complicated than that should be simplified before you try to solve. Before you try to solve equations with two terms on each side, however, let's start with a very simple equation that only takes one step to solve.

Every equation talks about what happened to an unknown number and what the result was. The equation $2y = 10$ says an unknown number y was multiplied by 2 and the result was 10. Solving the equation means figuring out the value of the unknown number. What number, multiplied by 2, gives you 10?

For simple equations, your basic number sense or a bit of trial and error can usually answer that question, but it's important to build a collection of skills that will get you to the answer when intuition isn't enough. To solve the equation, you start with the result, 10, and undo the multiplication that was done to the unknown by performing the opposite or inverse operation. To find out what value of the variable y solves the equation $2y = 10$, you can divide 10 by 2 to get back to $y = 5$. The variable was multiplied by 2 to get 10, so you divide 10 by 2 to get the value of the variable.

Multiplication and division are inverse operations. If the variable was multiplied by a number, you solve by dividing the result by that number. If the variable was divided by a number, multiply the result by that number. To solve $\frac{x}{5} = 12$, multiply 12 by 5, to find that $x = 60$. Addition and subtraction are inverses. Solve the equation $x + 7 = 22$ by subtracting 7 from 22 to get 15. Solve $t - 3 = 8$ by adding 3 to 8, for a solution of 11.

Now all that probably seems right. You can check that the answers are correct by asking "does 2 times 5 equal 10?" or "does 11 minus 3 equal 8?" But why does this "do the opposite" thinking work?

THINK ABOUT IT

Spending the time to understand why a particular process got you to the right answer guarantees that it wasn't just a happy accident. If you can explain why it works, you can be confident it will work all the time.

An equation is like a balanced scale. The expressions on both sides of the equal sign have the same value. If you make a change on only one side, you'll throw the scale out of balance. The expressions won't be equal anymore. But if you make the exact same change to both sides, the scale will remain balanced.

When you solve $x + 7 = 22$, you don't just subtract the 7 from the 22. You subtract 7 from both sides.

$$x + 7 = 22$$
$$x + 7 - 7 = 22 - 7$$

Doing the same thing on both sides keeps the scale balanced, and combining the like terms on each side gives you a simpler equation that tells you the value of the variable.

$$x + 7 = 22$$
$$x + 7 - 7 = 22 - 7$$
$$x = 15$$

Properties of the real numbers say that you can add the same number to both sides of an equation, subtract the same number from both sides, multiply or divide both sides by the same number, and the equation will stay balanced.

Addition Property of Equality	If $a = b$ then $a + c = b + c$
Subtraction Property of Equality	If $a = b$ then $a - c = b - c$
Multiplication Property of Equality	If $a = b$ then $a \cdot c = b \cdot c$
Division Property of Equality	If $a = b$ and $c \neq 0$ then $\dfrac{a}{c} = \dfrac{b}{c}$

 THINK ABOUT IT

The division property of equality has an extra bit that says the number you're dividing by can't be 0. Division by 0 is impossible. If you were asked to divide 12 by 3, you could think of it as breaking 12 into groups of 3, but if you try to divide 12 by 0, you can't break 12 into groups of nothing. While multiplying by 0 is legal, it's not useful either. It will give you $0 = 0$, no matter what you started with.

When you solve $\dfrac{x}{5} = 12$, this is what is really happening:

$$\frac{x}{5} = 12$$
$$\frac{x}{\cancel{5}} \cdot \frac{\cancel{5}}{1} = 12 \cdot 5$$
$$x = 60$$

To solve an equation, perform the inverse operation on both sides. That reduces one side to just the variable and the other to original value of the variable. Remember your goal is to get the variable all alone on one side and a number all alone on the other side. Focus on eliminating anything on the variable side that isn't the variable.

 CHECK POINT

Solve each equation by performing the inverse operation.

11. $5x = 35$

12. $t - 3 = 17$

13. $y + 18 = 45$

14. $16 = \dfrac{t}{3}$

15. $a + 39 = 14$

16. $w - 11 = 0$

17. $-8x = 72$

18. $t - 56 = 293$

19. $\dfrac{y}{15} = -8$

20. $z + 91 = 91$

Solving Two-Step Equations

To solve a one-step equation, you perform the opposite, or inverse, operation. Unfortunately, most equations you encounter involve more than one operation. The key will still be inverse operations, but which operation, and when?

To answer that question, you have to go back to the order of operations. An equation, like $3x - 5 = 16$, says that a certain number, x, was multiplied by 3, and then 5 was subtracted from the result, and the final answer was 16. Starting with what the number x represents, the order of operations tells you to multiply first, and then subtract. We want to do the inverse, or opposite. We want to undo what was done.

 THINK ABOUT IT

When you get dressed, you put on your socks, and then your shoes. When you undress, you take off your shoes before taking off your socks. Creating an expression is like getting the variable dressed. Solving an equation is undressing the variable. Inverse operations, opposite order.

Solving is going in the opposite direction, so you need to reverse the order of operations. The last thing that happened before arriving at the final answer of 16 was subtracting 5. Undo that part first by adding 5 to both sides of the equation.

$$3x - 5 = 16$$
$$3x - 5 + 5 = 16 + 5$$
$$3x = 21$$

Remember that an equation must be balanced. Each side of the equal sign must have the same value. In order to maintain that balance, you have to perform any operation you do to both sides.

Before that subtraction happened, but after the variable was multiplied by 3, you had 21. To get back to the original value of the variable, you need to undo that multiplication. You can do that by dividing both sides by 3.

$$3x - 5 = 16$$
$$3x - 5 + 5 = 16 + 5$$
$$3x = 21$$
$$\frac{\cancel{3}x}{\cancel{3}} = \frac{21}{3}$$
$$x = 7$$

To solve a two-step equation, follow these steps.

1. Identify the two operations that have been performed on the variable, and the order in which they were performed.

2. Perform the inverse of the last operation, and simplify.

3. Perform the inverse of the remaining operation.

 TIP

If you don't like working with fractions, you can multiply both sides of the equation by the common denominator of the fractions. That will eliminate all the fractions. It may give you bigger numbers to work with, so choose wisely.

To solve $-5x + 12 = 2$, first note that x was multiplied by -5, and then 12 was added. Undo the addition first by subtracting 12 from both sides.

$$-5x + 12 = 2$$
$$-5x + 12 - 12 = 2 - 12$$
$$-5x = -10$$

Undo the remaining multiplication by dividing both sides by -5.

$$-5x + 12 = 2$$
$$-5x + 12 - 12 = 2 - 12$$
$$-5x = -10$$
$$\frac{\cancel{-5}x}{\cancel{-5}} = \frac{-10}{-5}$$
$$x = 2$$

 CHECK POINT

Solve each equation by performing inverse operations in reverse order.

21. $3a + 5 = 26$

22. $2z - 7 = -15$

23. $-4n + 5 = 33$

24. $11y - 9 = 2$

25. $6t + 23 = -7$

26. $\frac{x}{2} - 19 = 21$

27. $3z - 2.5 = 5$

28. $-19w - 72 = 4$

29. $-\frac{2}{3}x + \frac{7}{3} = \frac{1}{3}$

30. $\frac{y}{7} + 31 = 0$

Equations with Variables on Both Sides

The simple skill you need to solve equations is do the opposite. Perform the inverse operation. If it's a two-step equation, perform the inverse operations in reverse order. The key to making those simple rules work for you is simplifying each side of the equation before you start solving. You never want to start solving until you have simplified enough that there are no more than two terms on either side.

Thus far, the equations we've solved all had one side with only one term. What if there are two terms on both sides? Or if the side that has only one term has a variable term, not a constant? Your strategy is still focused on inverse operations with just a small adjustment.

If the equation has variable terms on both sides, eliminate one by adding or subtracting. If you need to solve $7a - 3 = 5a + 9$, your first move is to eliminate one of the variable terms. You can choose either one to eliminate. Usually, we eliminate the smaller one, but it's entirely up to you. Let's eliminate the $5a$ by subtracting $5a$ from both sides.

$$7a - 3 = 5a + 9$$
$$7a - 3 - 5a = 5a + 9 - 5a$$
$$2a - 3 = 9$$

Once you've eliminated one of the variable terms, and there's only a constant on one side, then you can follow the plan for a two-step equation. Add 3 to both sides, then divide both sides by 2.

$$7a - 3 = 5a + 9$$
$$7a - 3 - 5a = 5a + 9 - 5a$$
$$2a - 3 = 9$$
$$2a - 3 + 3 = 9 + 3$$
$$2a = 12$$
$$\frac{\cancel{2}a}{\cancel{2}} = \frac{12}{2}$$
$$a = 6$$

If you find yourself with an equation that has two terms on one side and one term on the other, but it's a variable term, you can still choose to eliminate either variable term, but you'll do less work if you eliminate the variable term on the two-term side. Suppose you need to solve $8x - 9 = 5x$. You can eliminate the $5x$ or the $8x$. If you choose to eliminate the $5x$, it looks like this.

$$8x - 9 = 5x$$
$$8x - 9 - 5x = 5x - 5x$$
$$3x - 9 = 0$$
$$3x - 9 + 9 = 0 + 9$$
$$3x = 9$$
$$\frac{\cancel{3}x}{\cancel{3}} = \frac{9}{3}$$
$$x = 3$$

If you choose to eliminate the $8x$, it's a little shorter.

$$8x - 9 = 5x$$
$$8x - 9 - 8x = 5x - 8x$$
$$-9 = -3x$$
$$\frac{-9}{-3} = \frac{\cancel{-3}x}{\cancel{-3}}$$
$$3 = x$$

 TIP

There is often more than one correct way to solve an equation. You could, for example, solve without simplifying first. It's not "wrong." It just takes longer and has more opportunities to make mistakes. Have a plan, a habit, of how you do things, and you'll find your work is more accurate.

Some equations, like $x = x + 1$, have no solution. Sometimes you can see just by reading the equation that it makes no sense. A number can't equal more than itself. When you try to solve, all the variables disappear and what's left is a false statement.

$$x = x + 1$$
$$x - x = x + 1 - x$$
$$0 = 1$$

Zero clearly doesn't equal 1. The equation has no solution.

There are also equations for which all the variables disappear, but the resulting statement is true. When you try to solve $5x - 7 = 3(x - 2) + 2x - 1$, you find it simplifies to

$$5x - 7 = 3(x - 2) + 2x - 1$$
$$5x - 7 = 3x - 6 + 2x - 1$$
$$5x - 7 = 5x - 7$$

The fact that both sides are exactly the same tells you that any number could replace x and the equation would still balance. But if you didn't realize that and went on to solve, look at what will happen.

$$5x - 7 = 3(x - 2) + 2x - 1$$
$$5x - 7 = 3x - 6 + 2x - 1$$
$$5x - 7 = 5x - 7$$
$$5x - 7 - 5x = 5x - 7 - 5x$$
$$-7 = -7$$

All the variables have disappeared and the remaining statement is true. That's a sign that any real number will solve this equation. Equations that are true for any value of the variable are called identities.

 TIP

You can always check a solution by plugging the value you found back into the equation and simplifying both sides. If both sides come out the same, your solution is correct. If not, first double-check your work in the check, and if that's accurate, double-check your solution.

 CHECK POINT

Solve each equation by performing inverse operations in reverse order.

31. $3x - 7 = 5x + 1$

32. $2t + 5 = 3t - 2$

33. $\dfrac{y}{4} + 1 = y - 8$

34. $\dfrac{v}{3} - 8 = \dfrac{v}{5} - 2$

35. $11 + 8x = 12x - 17$

36. $-2(y - 7) = 4y - 6(y + 1)$

37. $8(x + 5) = 2(x - 3) + 7(x - 1)$

38. $2t - 3(t + 1) = (2 - t) - 5$

39. $(4z - 5) + (2z + 7) = (3z - 4) + 2(z + 3)$

40. $x - 13 = 5(x + 10) - 17$

You now have all the skills needed to solve first degree, or linear, equations in one variable. All equations with a single variable, with no exponents or roots, and no variables in denominators, can be solved by simplifying each side of the equation, and then using inverse operations. Those are powerful tools, but only if you know when and how to use them.

If someone asks you to build a model of the Eiffel Tower or the Empire State Building, you understand that they want you to construct a smaller, simpler version of the landmark. When someone asks you to do some mathematical modeling, they're suggesting that you use your math skills to write a clear compact mathematical sentence that describes what's going on in a particular situation.

Modeling with Linear Equations

Modeling is finding a way to represent a situation using math. Right now, we'll do that using algebra, specifically, linear equations. Sometimes the situation you're trying to represent is described in language that translates directly into algebraic representations.

Find two numbers such that one number is 12 more than another and their sum is 42. You can represent one number with a variable, say x, and the other as $x+12$. The word *sum* means addition, so $x+x+12=42$. Combine like terms on the left side of the equation, and $2x+12=42$. Then solve the equation.

$$2x+12=42$$
$$2x+12-12=42-12$$
$$2x=30$$
$$\frac{\cancel{2}x}{\cancel{2}}=\frac{30}{2}$$
$$x=15$$

Don't move on just yet, because you still need to put this solution in context. You were looking for two numbers, and you called one of them x. That number is actually 15. The other number is 12 more than that, so it must be 15 + 12, or 27. The two numbers are 15 and 27, and if you check, they do add to 42.

Much of the time, however, the situation you have to represent isn't presented in a way that translates quite so easily to equations. George has a budget of $40 for school supplies. He has already spent $10.05 on folders and pens, but still needs to buy notebooks. The type of notebook he prefers sells for $5.99. How many notebooks can George purchase?

There are a lot of distractions from the simple *this number plus that number equals the other number* language of the previous example. It may help to begin by trying to state the situation in simpler terms. You might state this one as what George spends on folders and pens plus what he spends on notebooks equals $40, or in even simpler terms, folders plus notebooks equals $40.

TIP

Take the time to define your variable. It helps you understand what the question is asking, it helps your reader understand the solution you're presenting, and it helps you reflect afterward on whether your answer makes sense. If n is the number of notebooks and you get $n = -32.4$, you know something went wrong.

Focus in on what you need to find out and define a variable. You need to find out how many notebooks George can buy, so let n = the number of notebooks. If George buys n notebooks, how much does he spend on notebooks? Each notebook costs $5.99, so he'll spend $5.99n$ on notebooks, and he's already spent $10.05 on folders and pens.

Organize the information.

Let n = the number of notebooks George can buy

Let $5.99n$ = the amount George spends on notebooks

Let 10.05 = the amount George spends on folders and pens

Let 40 = George's budget

State the situation simply.

(amount for notebooks) + (amount for folders and pens) = 40.

Insert the symbols.

$5.99n + 10.05 = 40$

Now you have an equation you can solve.

$$5.99n + 10.05 = 40$$
$$5.99n + 10.05 - 10.05 = 40 - 10.05$$
$$5.99n = 29.95$$
$$\frac{5.99n}{5.99} = \frac{29.95}{5.99}$$
$$n = 5$$

George can buy five notebooks. (Of course, even this isn't a "real life" situation. In real life, there'd be tax to consider, and the solution probably wouldn't be a tidy integer answer.)

When you're trying to model a situation with an equation:

- Read the problem carefully.

- State the situation in the simplest possible language.

- Identify what you need to find.

- Define a variable, and any variable terms that represent elements of the situation.

- Translate the simple statement of the situation into an equation.

- Solve the equation.

- Interpret the solution in the context of the situation, and make sure it makes sense.

CHECK POINT

For each problem, define a variable, write an equation that describes the situation, and solve the equation.

41. Six more than three times a number is twenty-seven. Find the number.

42. Twice a number increased by 4 is 8 more than the number. Find the number.

43. The sum of two consecutive numbers is 53. Find the numbers.

44. The cost of manufacturing storage containers is $100 plus $0.18 per container. If the total cost of an order was $1,000, how many storage containers were manufactured?

45. Francesca built a garden 3 feet longer than it is wide. She needed 30 feet of fencing to enclose the garden. Find the dimensions of the garden.

46. Christopher bought 3 hot dogs and spent $1.50 for a drink. His total bill was $12. Find the price of a hot dog.

47. Carlos saves $12 each week. Dahlia has already saved $270 and each week she spends $15 of that savings. When will they both have the same amount in savings?

48. Alyssa received a $100 gift certificate for her birthday and used it to order 12 e-books. Afterward there was $5.32 left on the gift certificate. What was the average price per e-book?

49. When her community organized a fundraiser, Sophie was asked to contribute packages of snacks that could be sold at the event. She decided to package her secret spiced almonds for sale. She bought almonds for $9 per pound, and spent $2 for spices. The total bill was $56. How many pounds of almonds did she buy?

50. Six times the difference of 12 and a number exceeds the sum of 12 and the number by 4. Find the number.

The Least You Need to Know

- Equations are sentences that say two expressions have the same value.
- Always simplify each side of an equation before trying to solve.
- To solve equations, perform the inverse operations in the opposite order.
- If there are variable terms on both sides of the equation, eliminate one by adding or subtracting.
- Some equations have no solution, and some are true for all values of the variable.

Inequalities

Like any language, algebra needs to be able to say different things. Expressions are the phrases of mathematical communication, and equations are one type of algebraic sentence. Equations say two expressions have the same value, but that's not always the case. Sometimes the values of the two expressions are unequal with one larger than the other. In this chapter, we'll look at statements of inequality, and see what information we can gain by solving them.

Types of Inequalities

Inequalities are algebraic sentences that tell you that one expression is larger or smaller than another. There are several different symbols that are used to communicate inequality. The simplest, but least commonly used, is the "not equal" sign: $\neq$. If you see $x \neq 5$, which is read "x is not equal to 5," you know that x is not the same as 5. That gives you some information, but not a lot, which probably explains why that symbol isn't used more.

In This Chapter

- The similarities and differences between equations and inequalities
- How to solve single inequalities
- Decomposing compound inequalities
- Representing solutions of inequalities on the number line

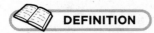 **DEFINITION**

An **inequality** compares two expressions that are not equal and shows which one is larger.

The more common inequality symbols are < for "is less than" and > for "is greater than." At times, you want to indicate that the upper or lower limit for the value of a variable is itself an acceptable value. In that case, you can add a small line segment that looks like part of an equal sign under the inequality sign. The symbol ≤ says "is less than or equal to" and the symbol ≥ says "is greater than or equal to."

This chart shows examples of the language that translates to inequalities.

x is less than 12.	$x < 12$
y is greater than 9.	$y > 9$
z exceeds 7.	$z > 7$
t is no more than 20.	$t \leq 20$
n is at most 13.	$n \leq 13$
w is no less than 15.	$w \geq 15$
a is at least 11.	$a \geq 11$

 CHECK POINT

Write an inequality for each sentence.

1. A number, x, is less than 12.

2. The interest rate, r, does not exceed 4%.

3. In order to ride the roller coaster, your height, h, must exceed 48 inches.

4. The price of a ticket, p, is at least $20.

5. Refunds are available for those whose income, I, is no more than $12,000.

Write a sentence for each inequality.

6. $x \geq 16$

7. $y < 12$

8. $t > 22$

9. $z \leq 100$

10. $a > 0$

Solving Inequalities

The solution of an inequality is not a single number. It's a range, or set, of numbers. Luckily, the process of finding that solution set is very similar to the process of solving an equation. In fact, there's only one significant difference.

To solve the inequality $3x - 5 < 2x + 12$, you can take the same steps you'd take to solve the equation $3x - 5 = 2x + 12$. You want to eliminate one of the variable terms by subtracting from both sides. $3x - 5 < 2x + 12$ becomes $3x - 5 - 2x < 2x + 12 - 2x$ or $x - 5 < 12$. Adding 5 to both sides tells you that $x < 17$. You now have a range of values for the variable.

 ALGEBRA TRAP

Unlike equations, inequalities are not symmetric. The equation $3x - 5 = 2x + 12$ can be rewritten as $2x + 12 = 3x - 5$, but $3x - 5 < 2x + 12$ and $2x + 12 < 3x - 5$ do not say the same thing at all. In fact, they're opposites. If you must rewrite an inequality, make sure that the smaller expression is at the smaller end of the inequality sign. $3x - 5 < 2x + 12$ is equivalent to $2x + 12 > 3x - 5$, not to $2x + 12 < 3x - 5$.

The solution to this inequality is not a single number. It's not 17, but it could be any number less than 17. You can test any number less than 17 to check. If I pick 2, $3 \cdot 2 - 5$ is 1, and $2 \cdot 2 + 12$ is 16, and 1 is less than 16. Replacing x with 2, or any number less than 17, will make the inequality statement true.

 TIP

Look carefully at the inequality sign. If the sign is < or > the number on the other side is not part of the solution set. If the sign has an "or equal to," as in ≤ or ≥, then the number is part of the solution set.

Let's look at an inequality for which you need to do something different. Take the inequality $-2x + 5 > 3$. Start solving it just as if it were an equation, by subtracting 5 from both sides.

$$-2x + 5 > 3$$
$$-2x + 5 - 5 > 3 - 5$$
$$-2x > -2$$

Now take a minute to play guess and test, and find a value for x that makes the statement true. You're dealing with negative numbers, so remember that you have to think about left and right on the number line. If you try $x = 2$, the inequality becomes $-2 \cdot 2 > -2$, which simplifies to $-4 > -2$, but that's not true. If you try $x = -2$, the inequality becomes $-2(-2) > -2$, which is $4 > -2$, a true statement. So $x = 2$ is not part of the solution but $x = -2$ is. Remember that and let's take another step in solving.

If you were solving an equation, you would divide both sides by -2. Let's try that.

$$\frac{-2x}{-2} > \frac{-2}{-2}$$
$$x > 1$$

That says that the solution is all numbers greater than 1, but you already know that 2, which is greater than 1, doesn't work and -2, which is less than 1, does. What went wrong?

When you divided by a negative number, you bumped into that "through the looking glass" appearance of negative numbers on the number line. When you solve an equation, you're looking for one number, so the mirroring doesn't matter. When you solve an inequality, the number you find is just a divider and you need to know which side of that border you want to be on. To get that straight you need to add one extra rule for inequalities: when you multiply or divide by a negative number, reverse the direction of the inequality sign.

The correct solution of $-2x + 5 > 3$ will look like this.

$$-2x + 5 > 3$$
$$-2x + 5 - 5 > 3 - 5 \qquad \text{Subtract 5 from both sides}$$
$$-2x > -2$$
$$\frac{-2x}{-2} > \frac{-2}{-2} \qquad \text{Divide both sides by -2}$$
$$x < 1 \qquad \text{Reverse inequality sign because of negative divisor}$$

To solve an inequality, follow these steps.

1. Simplify the left side and the right side so that there are no more than two terms on either side.

2. If there are variable terms on both sides, eliminate one of them by adding or subtracting the same term to both sides.

3. Isolate the variable by performing inverse operations.

4. If you multiply or divide both sides of the inequality by a negative number, reverse the direction of the inequality sign.

5. Graph the solution set on a number line.

6. Test a value from your solution set in the original inequality.

To solve the inequality $3(x-3)+1 \le 4(x-6)-2$, first simplify the left side.

$$3(x-3)+1 \le 4(x-6)-2$$
$$3x-9+1 \le 4(x-6)-2$$
$$3x-8 \le 4(x-6)-2$$

Then simplify the right side.

$$3x-8 \le 4(x-6)-2$$
$$3x-8 \le 4x-24-2$$
$$3x-8 \le 4x-26$$

Eliminate one of the variable terms by subtracting $4x$ from both sides.

$$3x-8 \le 4x-26$$
$$3x-8-4x \le 4x-26-4x$$
$$-x-8 \le -26$$

Add 8 to both sides.

$$-x-8 \le -26$$
$$-x-8+8 \le -26+8$$
$$-x \le -18$$

Divide both sides by -1 and reverse the inequality sign.

$$-x \leq -18$$

$$\frac{-x}{-1} \leq \frac{-18}{-1}$$

$$x \geq 18$$

Graph the solution on a number line.

Test a value from the solution set in the original inequality. Pick a number larger than 18, like 20, and substitute that for x in the original inequality.

$$3(x-3)+1 \leq 4(x-6)-2$$

$$3(20-3)+1 \leq 4(20-6)-2$$

$$3(17)+1 \leq 4(14)-2$$

$$51+1 \leq 56-2$$

$$52 \leq 54$$

The result after simplifying is a true statement—52 is truly less than 54—so you can be confident that your solution is correct.

  CHECK POINT

Solve each inequality.

11. $x+7 > 4$

12. $y-7 < -1$

13. $2x-3 \geq 5$

14. $5-3t < 23+t$

15. $3y+11 \leq 17-y$

For each inequality, tell whether the indicated value of the variable is part of the solution.

16. $-7x-9 \geq 12$, $x = -4$

17. $5-x > 11$, $x = -6$

18. $2y+3 < 4$, $y = 1$

19. $6-3x \leq -12$, $x = 5$

20. $11t+19 \leq 8t+61$, $t = 10$

Picturing Solutions on a Number Line

Because the solution of an inequality is not a single number but a whole set of numbers, it's helpful to have a visual representation of the solution. You can create that picture by graphing the solution set on a number line.

To graph the solution of an inequality on the number line, you first need to locate that number that divides the numbers that are part of the solution from those that are not. If your solution is $x \geq -3$, -3 is the dividing point.

 TIP

Plan ahead when graphing the solution of an inequality on a number line. The number line goes on forever, so you can't show all of it, and you can choose what part and how much you want to show. Your tick marks and labels don't need to count by ones either. If you're graphing $x \geq 150$, you don't want to start at 0 and count by ones. Choose the range and scale that make sense for your task.

Before you mark that number, you must decide whether that border number is or is not part of the solution. If the inequality is $\geq$ or $\leq$, the number that divides the line is part of the solution. Use a solid dot to mark that point.

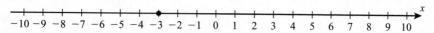

If the symbol is $<$ or $>$, the number is not part of the solution. You still need to mark the number, but do it by just putting an open circle. If you wanted to show $x > -3$ instead of $x \geq -3$, you'd mark -3 like this.

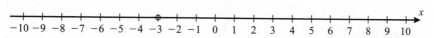

Once you've marked the dividing point, you have to show whether the solution falls above or below that. You do this by shading the side that is the solution, and placing an arrow on the line of shading to show that it goes on forever. For $x \geq -3$, it looks like this.

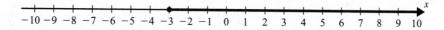

CHECK POINT

Graph each inequality on a number line.

21. $x > 5$ 24. $z \geq 7$

22. $t \geq -8$ 25. $a \leq 9$

23. $y < -4$

Solve each inequality and graph the solution on a number line.

26. $3x - 7 \leq 5$ 29. $-6y + 5 < 3y + 14$

27. $x - 6 > 5x + 2$ 30. $5(t-2) \leq (2t-3) + (t-7)$

28. $-2x + 1 \geq 9$

Compound Inequalities

Solving an inequality is just like solving an equation except that you reverse the inequality sign if you multiply or divide both sides by a negative number. The solution to an inequality is a set of numbers, all the numbers above or below a certain boundary point. Sometimes the boundary is part of the solution and sometimes it is not.

Having such a large set of solutions can be unsatisfying. Although you can't get a specific number, you can narrow things down. *Compound inequalities* are one way we can do that. A compound inequality is a statement made up of two inequalities connected either with the word *or* or the word *and*.

DEFINITION

A **compound inequality** is actually two inequalities connected by either the word *and* or the word *or*. If the *and* is used, the compound inequality is called a conjunction. If the pieces are joined by *or*, it's called a disjunction.

A disjunction, or *or* inequality, might look like $7x + 5 \leq -16$ or $11 - 3x < -1$. It truly is just two inequalities, using the same variable, connected by the word "or." To solve the compound statement, solve each of the individual inequalities and connect the solutions with an "or." (Don't forget to reverse the inequality sign if you divide by a negative.)

$$7x + 5 \leq -16 \quad \text{or} \quad 11 - 3x < -1$$
$$7x \leq -21 \quad \text{or} \quad -3x < -12$$
$$x \leq -3 \quad \text{or} \quad x > 4$$

Now you know that the numbers that meet the description in this compound statement are -3, all the numbers less than -3 and all the numbers greater than 4. Turning that around, you could say that the numbers that are not in the solution are numbers greater than -3 and less than or equal to 4. Here's what it looks like on the number line.

The solution to an *or* inequality usually looks like two arrows pointing away from one another, with a space between them. If you find that the two arrows overlap, the solution is all real numbers.

The *or* inequality gives a little more information about the solution, but usually leaves us with a very large set. The *and* inequality generally restricts the solution set more effectively, but it takes an extra step or two to solve. Consider the compound inequality $7 \leq 2x - 3 < 11$. The *and* inequality is often presented in this condensed form. You don't see the word *and*, as the two inequalities are compressed into what looks like one and a half.

The inequality $7 \leq 2x - 3 < 11$ is actually $7 \leq 2x - 3$ and $2x - 3 < 11$, two inequalities, with the same variable, connected by the word "and." What allows them to be compressed is that they share the expression $2x - 3$ and both inequality signs point in the same direction. The statement $7 \leq 2x - 3 < 11$ says that $2x - 3$ is between 7 and 11. It's at least 7, but less than 11.

To solve a compound inequality like this one, first break it into its two inequalities. The first inequality starts at the beginning of the statement, and stops just before the second inequality sign. $\overset{\text{First Inequality}}{\overline{7 \leq 2x - 3}} < 11$. The second inequality begins just after the first inequality sign and goes to the end of the statement. $7 \leq \overset{\text{Second Inequality}}{\overline{2x - 3 < 11}}$. Connect the inequalities with the word "and" and then solve each separately. After you've solved each inequality, you usually can compress the two solutions into a single statement.

$$7 \leq 2x - 3 < 11$$

$7 \leq 2x - 3$	and	$2x - 3 < 11$
$10 \leq 2x$	and	$2x < 14$
$5 \leq x$	and	$x < 7$

$$5 \leq x < 7$$

The graph of this solution set has a solid dot at 5, an open circle at 7, and shades all the numbers in between.

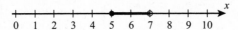

Don't try to force the compression of compound inequalities. They must be *and* inequalities, the inequality signs must point in the same direction, and there must be a shared expression that fits in the middle. You can compress $5 < 3x+1$ and $3x+1 \leq 12$ into $5 < 3x+1 \leq 12$, but you can't compress $5 > 3x+1$ with $3x+1 \leq 12$, or $3x+1 < 5$ with $3x+1 \leq 12$.

 THINK ABOUT IT

The solution of an *and* inequality can usually be condensed, if the original was condensed. Your two separate solutions will have the form $a < x$ and $x < b$, with $a < b$. If $a > b$, however, you have a contradiction. It's possible to have $3 < x < 5$ but it's not possible to have $5 < x < 3$. If x is greater than 5, it can't be less than 3. If you find yourself in this situation, first check to see if you divided by a negative and forgot to reverse the inequality sign. If there's no mistake, the inequality has no solution.

The compound inequality $-5 \leq 7-3x \leq 43$ can be separated into two inequalities connected by an *and*.

$$-5 \leq 7-3x \leq 43$$

$-5 \leq 7-3x$	and	$7-3x \leq 43$
$-5-7 \leq 7-3x-7$	and	$7-3x-7 \leq 43-7$
$-12 \leq -3x$	and	$-3x \leq 36$
$\dfrac{-12}{-3} \leq \dfrac{-3x}{-3}$	and	$\dfrac{-3x}{-3} \leq \dfrac{36}{-3}$
$4 \geq x$	and	$x \geq -12$

$$4 \geq x \geq -12$$
$$-12 \leq x \leq 4$$

Dividing both sides by -3 caused the inequality sign to change from $\leq$ to $\geq$. The solution, presented from large to small, $4 \geq x \geq -12$, is correct and acceptable, but you might want to carefully rewrite it from small to large, $-12 \leq x \leq 4$, just because that's the more familiar form. Either way, it looks like this when graphed.

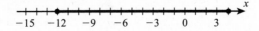

CHECK POINT

Solve each compound inequality. Graph the solution set on the number line.

31. $4y - 7 > 5$ or $5y + 7 < -3$

32. $-1 \leq 6z - 7 \leq 11$

33. $3x - 8 \leq 4x - 26$ or $x + 5 < 12$

34. $2 < 3t - 7 < 5$

35. $17 < 4y + 5 \leq 1$

36. $7t - 3 \leq 8t + 5$ or $4t + 1 > 2t - 9$

37. $-11 \leq 3x + 1 \leq 19$

38. $9 < 2x - 1$ or $2x - 1 < 6$

39. $21 + t < 5 - 3t \leq 20$

40. $y - 2 \leq 5(y - 2) < (y - 2) + 8$

The Least You Need to Know

- Solve inequalities as if they are equations, but reverse the inequality sign if you multiply or divide by a negative number.

- Compound inequalities are two inequalities connected by "or" or "and." Solve each of the inequalities separately.

- Graph the solution set of an inequality on a number line.

Graphing Linear Functions

A picture is worth a thousand words, according to the old saying, and pictures can be just as valuable in helping us understand equations and functions. The pictures that represent equations and functions tend to be one picture for one equation rather than one picture for a thousand of anything, but these pictures, or graphs, make patterns clearer.

We've already talked a little bit about graphs when we first looked at functions, and we saw that the graph gave us a quick way to tell if a relation was a function. In this chapter, we'll focus on the graphs of an important group of functions, called the linear functions. We'll see why they have that name, and we'll look at quick graphing methods and one important application.

In This Chapter

- Understanding the coordinate plane
- Graphing using a table of values
- Quick graphing by intercept-intercept or slope-intercept
- Understanding direct variation relationships

The Cartesian Plane

The system of graphing that you're probably accustomed to using is a rectangular coordinate system, or *coordinate plane*. It gets its formal name, the *Cartesian coordinate system*, from Rene Descartes, a seventeenth-century French mathematician and philosopher. Using this system, you locate each point by a pair of numbers, called an ordered pair because the order of the numbers is important. The first number tells you how many spaces to move left or right, and the second number indicates movement up and down.

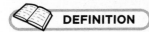

DEFINITION

The **Cartesian plane**, or **coordinate plane**, is a system of identifying every point in the plane by an ordered pair of numbers. The plane is divided into four sections, called **quadrants**, by a horizontal line, called the x-axis and a vertical line called the y-axis, which intersect at the point called the **origin**. The first number in the ordered pair tells how to move left or right from the origin and the second number tells how to move up or down.

All this movement starts from a point called the *origin*, which is the intersection of a horizontal line and a vertical line. The horizontal line is generally called the x-axis and the vertical line is the y-axis. The origin is the point (0, 0). On the x-axis, positive numbers go to the right and the negatives to the left, just like a number line. On the y-axis, positives go up and negatives down.

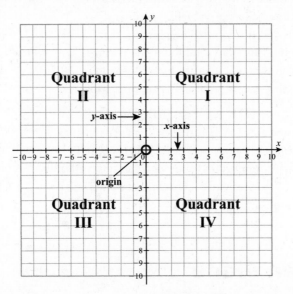

The x-axis and y-axis intersect at the origin and divide the plane into four sections called *quadrants*. The quadrants are referred to by numbers, starting from the upper right and going counterclockwise.

THINK ABOUT IT

The graphing we're talking about draws pictures on a flat surface, or plane. You can think of a plane as a piece of paper that never ends, but it's basically two-dimensional, which is why two coordinates are all you need to locate a point. If you were working in three dimensions, you'd need three axes, and three coordinates. The third number would tell you to move up above the page or down below it.

 CHECK POINT

Plot each point on the Cartesian plane.

1. (-4, 7)

2. (6, -2)

3. (0, 4)

4. (-3, -1)

5. (-2, 0)

Use the figure to answer questions 6 through 10.

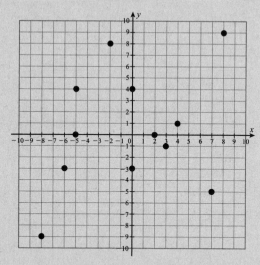

6. Give the coordinates of a point in the first quadrant.

7. Give the coordinates of a point on the *y*-axis.

8. Give the coordinates of a point in Quadrant IV.

9. Give the coordinates of a point in Quadrant III.

10. Give the coordinates of a point on the *x*-axis.

Graphing Linear Functions

In an earlier chapter on functions, we talked about graphing pairs of numbers as points in the plane to help decide if a relation is a function, and if it follows a pattern. One of the most common patterns you will see is a line. Functions whose graphs form a line are called *linear functions.* The pattern can be represented by an equation, which can be written using function notation. Throughout this chapter, we'll use "linear function" and "linear equation" interchangeably.

 **DEFINITION**

A **linear function** is a function of the form $y = mx + b$. It defines the value of the output variable as a multiple of the input variable, possibly plus or minus a constant.

You can recognize a linear function by looking at its rule. The rule has a term that includes the input or independent variable, raised to the first power. You won't see any exponents or roots. If there's a fraction, the denominator contains only numbers, never variables. There may or may not also be a constant term. $f(x) = -7x$ is a linear function, and so is $g(x) = \dfrac{x}{2} + 3$, but $y = \dfrac{2}{x}$ and $p(x) = x^2 - 4x + 6$ are not linear functions.

Table of Values

The most basic way to graph a linear function—or any function, for that matter—is to use the rule to build a table of values. If the rule for your function is $f(x) = 2x - 5$, you can choose several values for the independent variable, x, to build a table. It's wise to choose both positive and negative values for x. You're not saying these values are the entire domain of the function. You're just using them to give an idea of what the graph looks like.

x	-4	-3	-2	-1	0	1	2	3	4
$f(x)$									

Putting the values you've chosen in order makes them easier to plot, but there's no rule that says you have to choose consecutive numbers for your table. If there are fractions involved in the rule you're going to evaluate, you can choose numbers that are divisible by the denominator. If you need to multiply your inputs by $\dfrac{1}{3}$, for example, you can choose to use only multiples of three to eliminate fractions.

Once you've chosen the values for the independent variable, you evaluate $2x - 5$ for each value.

x	-4	-3	-2	-1	0	1	2	3	4
$f(x)$	$2 \cdot (-4) - 5$ $= -13$	$2 \cdot (-3) - 5$ $= -11$	$2 \cdot (-2) - 5$ $= -9$	$2 \cdot (-1) - 5$ $= -7$	$2 \cdot 0 - 5$ $= -5$	$2 \cdot 1 - 5$ $= -3$	$2 \cdot 2 - 5$ $= -1$	$2 \cdot 3 - 5$ $= 1$	$2 \cdot 4 - 5$ $= 3$

The table of values now gives you a set of points that belong to the function. It's not the whole function. You're assuming the function has a *domain* of all real numbers. These points should be enough to show you the pattern, however, so plotting them will be the next step.

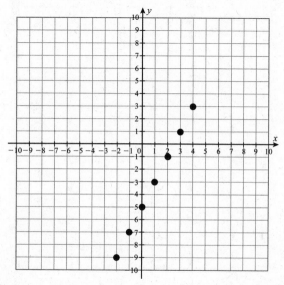

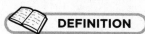 **DEFINITION**

The **domain** of a function is the set of all values that can be substituted for x, all possible inputs. The **range** is the set of all outputs, all values of y.

Once the pattern is clear, you just need a straightedge to draw the line. Add arrowheads on the ends of the line to communicate the fact that it continues beyond what you've drawn.

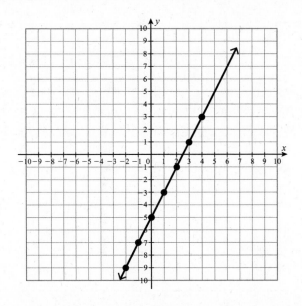

CHECK POINT

Graph each function by making a table of values and plotting points.

11. $f(x) = 3x - 4$

12. $g(x) = \dfrac{1}{2}x + 4$

13. $y = 2x - 3$

14. $g(x) = x + 5$

15. $y = -x + 5$

16. $f(x) = -2x + 7$

17. $g(x) = -\dfrac{2}{3}x + 6$

18. $y = 4 - 3x$

19. $f(x) = -2x$

20. $g(x) = -\dfrac{3}{2}x + 5$

After building a few tables of values, you may have begun to notice some patterns, or perhaps patterns in the patterns. Those are going to lead to some shortcuts for graphing linear functions without having to build the whole table of values. You probably noticed that whenever you chose zero as a value of x, the arithmetic was easy. That's at the heart of the first shortcut.

Intercepts

The x-axis and y-axis are the backbone of the coordinate system. The point at which a graph crosses an axis is called an intercept. The graph crosses the x-axis at the x-intercept, and crosses the y-axis at the y-intercept. Intercepts may not seem especially important. They are just points of the graph that happen to sit on an axis. But geometry dictates that two points are enough to define a line, and every point on an axis has a particular trait in common: one of its coordinates is 0. The x-intercept of a graph is a point that has a y-coordinate of 0. The y-intercept of a graph is a point that has an x-coordinate of 0. Zeros make for easy arithmetic, and quickly give you two points to define the line.

ALGEBRA TRAP

The intercept-intercept method won't work if the y-intercept is 0, because the x-intercept will also be 0. Both intercepts are at the origin, so you only get one point instead of two. You can't tell where the line goes with only one point. You'll need to use a different method.

To graph a linear function like $f(x) = 3x - 9$, begin by writing it as $y = 3x - 9$. Find the y-intercept by replacing x with 0. $y = 3 \cdot 0 - 9 = -9$. The y-intercept is $(0, -9)$. Then find the x-intercept by replacing y with 0. This may require a little bit of algebra, but it shouldn't be too tough.

$$0 = 3x - 9$$
$$9 = 3x$$
$$3 = x$$

The *x*-intercept is (3, 0). If you plot (0, −9) and (3, 0), you can use a straightedge to connect them into a line.

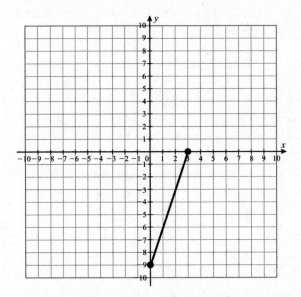

Extend the line and add arrowheads on the ends, and you have the graph of $y = 3x - 9$.

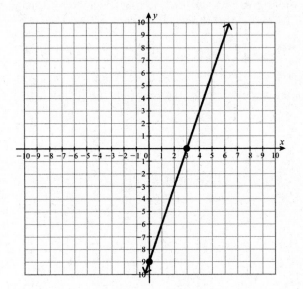

CHECK POINT

Find the x-intercept and y-intercept of each function.

21. $y = 2x - 6$ 24. $y = 5 - x$

22. $y = -3x + 6$ 25. $y = x + 4$

23. $y = 4x + 4$

Graph each function using intercepts.

26. $y = -2x + 8$ 29. $y = x + 5$

27. $y = 3x - 6$ 30. $y = \dfrac{1}{2}x - 4$

28. $y = x - 6$

When you built tables of values, you noticed that letting x equal 0 left you with easy arithmetic. That led to the intercept-intercept shortcut. When you were working with intercepts, did you notice there was a way to spot the y-intercept quickly? That ability to spot a y-intercept quickly is going to take you to another shortcut.

Slope and y-intercept

As you worked with the intercept-intercept method, you may have noticed that if the equation of the linear function was in the form $y =$ variable term + constant, that constant term was always the y-intercept. You needed to do some math to find the x-intercept, but the y-intercept was just sitting there.

 TIP

If you don't see a constant term, the y-intercept is 0 and the line passes through the origin.

You might have noticed something else surprising when you made tables of values. If you picked inputs in sequence, the change in the output as you went from one to the next was equal to the coefficient of the variable term. If you made a table for $y = 2x + 1$ and chose your x values that were consecutive, like 1, 2, 3, 4, and 5, the y values went up by twos.

x	1	2	3	4	5
$y = 2x + 1$	3	5	7	9	11

If you change the function rule to $y = 3x + 1$, the increase in y at each step will be 3.

x	1	2	3	4	5
$y = 3x + 1$	4	7	10	13	16

That constant rate of change you're noticing is called the *slope* of the line. If the coefficient of the variable term is positive, the line will rise from left to right. If the coefficient is negative, the line will fall. A horizontal line, which neither rises nor falls, has a slope of zero. The larger the absolute value of the slope, the steeper the rise or fall of the line.

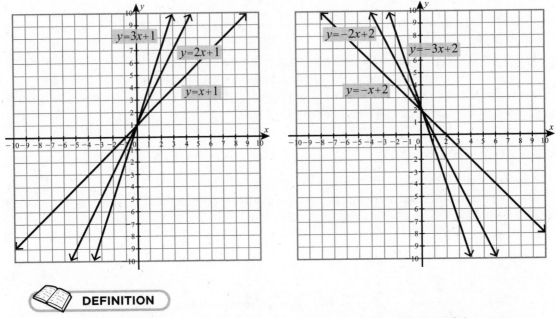

DEFINITION

The **slope** of the graph of a linear equation is the coefficient of the x-term, and tells how much the line rises or falls each time you move one unit right.

Once you understand what these two numbers, the slope and the y-intercept, tell you about the graph of the function, you can use them to draw the graph quickly. To graph $y = \frac{1}{2}x + 3$, first identify the y-intercept. In this equation, that's the constant term, 3, or (0, 3). Plot the point (0, 3) on the y-axis.

Next, focus on the slope, in this case, $\frac{1}{2}$. That tells you that each time you increase the x-coordinate by 1, you increase the y-coordinate by $\frac{1}{2}$. Because most graph paper isn't marked in half boxes, this could be tricky, so instead of counting over 1 and up $\frac{1}{2}$, count over 2 and up 1. That's still equivalent to $\frac{1}{2}$, but gives you a convenient way to think about the slope. The denominator tells you how much to move over, and the numerator tells how much to move up or down. This is why we talk about the slope as the ratio $\frac{\text{rise}}{\text{run}}$.

Starting from the y-intercept, count 2 units to the right and up 1. Plot a point. Do it again. From where you just placed the point, count 2 units right and 1 up and plot a point. After you've done that a few times, you'll be able to connect the points into a line.

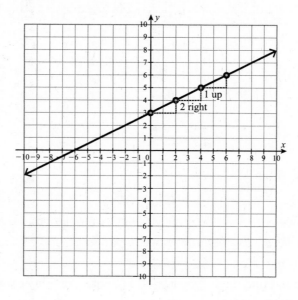

If the slope of the linear function is a negative number, count right and down. For the line $y = -2x + 6$, start at the y-intercept (0, 6), make the slope of -2 a fraction by writing it as $\frac{-2}{1}$, and then count over 1 and down 2, over 1 and down 2. Connect the dots in a line.

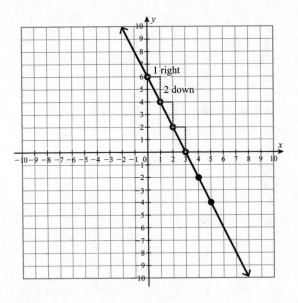

![Algebra Trap] **ALGEBRA TRAP**

If the slope is negative, be careful. You can count it as $\frac{\text{down}}{\text{right}}$, down is a negative direction, and right is a positive direction. A negative over a positive gives you a negative slope. If you must, you can count it as $\frac{\text{up}}{\text{left}}$, a positive direction over a negative direction. But don't get carried away. If you go left and down, you'll have a negative over a negative and wind up with a positive slope.

We say that slope $= \frac{\text{rise}}{\text{run}}$ but when the slope is negative, that's really $\frac{\text{fall}}{\text{run}}$ if you're moving from left to right. Count the run moving to the right, and let the sign of the slope tell you whether to rise or fall.

The combination of the y-intercept and the slope, both of which are easily identified just by looking at the equation, allows you to draw the graph quickly by following these steps.

1. Identify the y-intercept and plot that point.

2. Identify the slope. If the slope is a whole number, make it a fraction by giving it a denominator of 1.

3. From the y-intercept, count up or down the number in the numerator of the slope and over to the right the number in the denominator. Place a dot.

4. Repeat several times. Connect the dots into a line.

TIP

Sometimes when you're graphing, you may find that you just can't count the slope from right to left, usually because going in that direction will run off the page. You have to go "backward" or to the left. That reverses your run, so you need to reverse your rise or fall as well.

$$\frac{up}{right} = \frac{down}{left}$$

CHECK POINT

Graph each function by slope and y-intercept.

31. $y = 2x - 5$

32. $y = 3x - 7$

33. $y = -4x + 9$

34. $y = \frac{1}{2}x - 3$

35. $y = \frac{2}{3}x - 4$

36. $y = -\frac{4}{3}x + 7$

37. $y = -x + 5$

38. $y = 2x - 9$

39. $y = 2x + 3$

40. $y = -2x + 1$

There are a great many situations that can be modeled by linear functions, but one in particular is very common and happily, very simple.

Direct Variation

When two quantities are related and one changes, the other changes as well. If you buy two doughnuts, you might pay $1.78. If you decide you're hungry and add a third doughnut to your order, the price will increase to $2.67. The price varies with the number of doughnuts you buy.

There are different kinds of variation, but the most common one, direct variation, is modeled by a simple linear equation. When two quantities vary directly, one is a constant multiple of the other. If muffins are $2 each, the price of an order of muffins is $2 times the number of muffins.

The cost, C, of the order varies directly with the number of muffins, m, and you can describe that with the equation $C = 2m$. Direct variation relationships always fit that simple equation. There is no (non-zero) constant term, and nothing going on except one variable being multiplied by a constant to produce the other.

The constant multiplier is called the constant of variation, and it is usually denoted as k. The general pattern of a direct variation equation is $y = kx$. The specific equation of a direct variation relationship has that form, but uses the particular variables from the relationship and fills in the value of k. $C = 2m$ is the particular equation of our muffin example. That's a linear function with a slope of 2 and a y-intercept of 0. In any direct variation relationship, the equation is a linear equation with a slope of k and a y-intercept of 0.

 THINK ABOUT IT

If y varies directly with x, it's also true that x varies directly with y. If $y = 2x$, $x = \dfrac{1}{2}y$. Both are direct variation, but the constant of variation is different.

Do you remember that doughnut example? If you buy 2 doughnuts, you pay $1.78, but if you buy 3, you pay $2.67. It's a direct variation relationship, but what's k? Let's say n is the number of doughnuts and P is the price. P varies directly with n, so $P = kn$. You don't know the value of k, but you can find it. Plug in what you do know: when $n = 2$, $P = 1.78$ or when $n = 3$, $P = 2.67$. $P = kn$ becomes $1.78 = k \cdot 2$ and dividing both sides by 2 tells you that $k = \dfrac{1.78}{2} = 0.89$. The particular equation for the doughnut example is $P = 0.89n$. You could use that equation to figure out that 6 doughnuts would cost $P = 0.89 \cdot 6 = \$5.34$.

CHECK POINT

In questions 41 through 45, tell whether the relationship is a direct variation relationship.

41. $y = 3x$

42. $y = 5x - 3$

43. $xy = 12$

44.

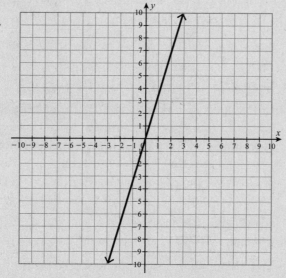

45.

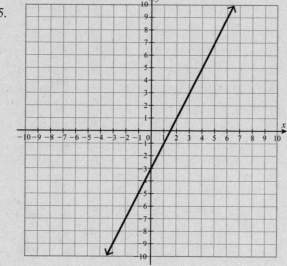

46. Distance traveled, d, and time spent traveling, t, vary directly. If you travel 90 miles in 2 hours, what is the constant of variation?

47. The circumference, C, of a circle varies directly with its radius, r. If $C = 20\pi$ when $r = 10$, what is the constant of variation?

48. The force, F, applied to an object, and the object's acceleration, a, are directly related. Force is measured in Newtons and acceleration in meters per second. A force of 200 Newtons applied to a certain object results in an acceleration of 4 meters per second. Write the equation that defines this particular relationship.

49. The cost of an order of potatoes, C, varies directly with the number of pounds, p, that are purchased. If 5 pounds of potatoes cost $7.45, how much will you pay for 2 pounds?

50. The volume, V, of a gas under constant pressure varies directly with its temperature, T. If a certain gas has a volume of 1,200 cubic centimeters when its temperature is 20°C, what will the volume be when the temperature is increased to 30°C?

The Least You Need to Know

- Linear functions can be described by equations of the form $y = mx + b$ where m is the slope and b is the y-intercept.

- The slope of a line is a number that describes how much the line rises or falls each time it moves one unit to the right.
  $$\text{slope} = \frac{\text{rise}}{\text{run}}$$

- The y-intercept is the point at which the line crosses the y-axis, where the x-coordinate is 0. The x-intercept is the point at which the line crosses the x-axis, where the y-coordinate is 0.

- You can graph a linear function by making a table of values, plotting points and connecting, or by finding both intercepts, plotting them and connecting, or by plotting the y-intercept, using the slope to plot more points, and connecting.

- Direct variation relationships fit a simple linear model $y = kx$ where k is the constant of variation and the y-intercept is always zero.

The Equation of a Line

Linear equations and functions are an important part of algebra. Not only do linear functions model many different situations on their own, but we often deal with other types of functions by translating them into linear functions. It makes sense to know how to solve linear equations and graph linear functions. Sometimes, however, we find ourselves needing to work in a different direction.

Many times we have information from which we can construct a graph, and when the graph is done, we realize we're looking at a linear relationship. What we need to figure out is the equation that describes that linear function. In this chapter, we'll look at the different forms in which a linear equation can appear, and how we take the information we have and work our way to the equation of the linear relationship.

In This Chapter

- Converting from one form of a linear equation to another
- Understanding the advantages of each form
- Finding the equation of a line
- Finding the equation of a line parallel to or perpendicular to a given line

Forms of the Linear Equation

When you first looked at linear equations and functions, you learned that a linear function has a simple variable term and a constant term. Of course, things aren't always that simple and straightforward. When written in function form, using $f(x)$ notation, linear functions usually appear just that way, $f(x) = mx + b$. When you use y or another variable to describe the relationship, things can get rearranged and look a bit different. Sometimes that's useful, but other times you just want to get back to that $y = mx + b$ form.

THINK ABOUT IT

What's the difference between a linear equation and a linear function? Linear *equations*, like $3x - 12 = 9$, have one variable, with no exponents, and need to be solved by finding the value of the variable. Linear *functions* connect two variables, usually x and y, by showing how they're related or defining one in terms of the other, like $y = 2x - 7$. The functions are the ones you'll graph.

You can always check whether an equation is linear by checking on what's not there. Linear equations won't have exponents, and won't have variables under radicals or in denominators. But you also want to get to know the three different forms in which we see linear functions presented.

Slope-Intercept Form

Slope-intercept form, commonly referred to as $y = mx + b$ form, should be familiar to you now. The form takes its name from the fact that it lets you see at a glance the slope of the line, m, and its y-intercept, b. This makes it a favorite form for graphing. If you have to graph $y = 2x - 7$, you know immediately that the y-intercept is (0, -7) and the slope is $\frac{2}{1}$. With that information, you can quickly sketch the graph.

DEFINITION

Slope-intercept form defines the output variable, y, in terms of a multiple of the input variable, x, and a constant, b. $y = mx + b$

If you're facing an equation that isn't in slope-intercept form, you can rearrange it by the same kind of algebraic operations that you use to solve an equation. In this case, you want to solve for y, or isolate y. The equation $5y - 3x - 3 = -9$ is not in slope-intercept form, but if we can get the y all alone on one side, we can get it to slope-intercept form. To do that, follow these steps.

1. Move any term that does not contain a y to the side opposite the y term by adding or subtracting. You can do this in one step or several. Expect to end up with the y term on one side, and up to two terms, an x term and possibly a constant term, on the other.

$$5y + 3x - 3 = 9$$
$$5y + 3x - 3 + 3 = 9 + 3$$
$$5y + 3x = 12$$
$$5y + 3x - 3x = 12 - 3x$$
$$5y = 12 - 3x$$

2. Rearrange the x-term and constant term, if necessary. The commutative property lets you do this. Remember you can think of subtraction as adding a negative.

$$5y = 12 - 3x$$
$$5y = 12 + -3x$$
$$5y = -3x + 12$$

3. Divide both sides by the coefficient of y.

$$5y = -3x + 12$$
$$\frac{5y}{5} = \frac{-3x + 12}{5}$$
$$y = \frac{-3x + 12}{5}$$

4. Simplify by dividing each term by the divisor. If the results of the division are not integers, fractions are more useful than decimals for slopes.

$$y = \frac{-3x + 12}{5}$$
$$y = -\frac{3}{5}x + \frac{12}{5}$$

 CHECK POINT

Put each equation in slope-intercept form.

1. $3x + 2y = 6$
2. $6y - 3 = 12x$
3. $x + 5y = 10$
4. $3x - y = 9$

5. $4 - 2y = 8x$
6. $4x - 8y = 16$
7. $2y + 12x - 6 = 0$
8. $\frac{1}{2}x - \frac{1}{3}y = \frac{5}{6}$

9. $4.5x + 3.5y + 31.5 = 0$
10. $\frac{y}{5} - 2 = 2x$

Standard Form

Another form in which you often see linear equations is called *standard form*. Standard form has an x term and a y term on the same side, and a constant on the other. This is the form you'd produce if you wanted to say that 4 hot dogs and 2 sodas cost $9. You'd write $4x + 2y = 9$. The x term and y term describing what you bought are added on the left side and the total cost is on the right side.

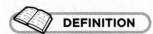 **DEFINITION**

Standard form is a way of writing an equation in which all the variable terms are on the same side, all the coefficients are integers, and the first coefficient is positive. $ax + by = c$

To put an equation in standard form, the basic step is to use addition or subtraction to move the variable terms to one side and the constant term to the other. There is more to the definition of standard form, however, that may add a step or two. Let's look at the equation $-\dfrac{3}{2}y = \dfrac{4}{5}x + 7$ and put it in standard form.

1. Get the variables on one side and the constant on the other by moving the x term to the side with the y term.

$$-\frac{3}{2}y = \frac{4}{5}x + 7$$

$$-\frac{3}{2}y - \frac{4}{5}x = \frac{4}{5}x + 7 - \frac{4}{5}x$$

$$-\frac{3}{2}y - \frac{4}{5}x = 7$$

2. Rearrange the left side so the x term comes first.

$$-\frac{3}{2}y - \frac{4}{5}x = 7$$

$$-\frac{4}{5}x - \frac{3}{2}y = 7$$

3. Change the coefficients to integers by multiplying both sides of the equation by the common denominator of all the fractions. In this case that's $5 \cdot 2 = 10$.

$$-\frac{4}{5}x - \frac{3}{2}y = 7$$

$$10\left(-\frac{4}{5}x - \frac{3}{2}y\right) = 7 \cdot 10$$

$$\frac{\overset{2}{\cancel{10}}}{1}\cdot\left(-\frac{4}{\cancel{5}}x\right) + \frac{\overset{5}{\cancel{10}}}{1}\cdot\left(-\frac{3}{\cancel{2}}y\right) = 70$$

$$-8x - 15y = 70$$

4. Make the first coefficient positive by multiplying both sides by -1.

$$-8x - 15y = 70$$

$$-1(-8x - 15y) = 70 \cdot (-1)$$

$$-1(-8x) + (-1)(-15y) = -70$$

$$8x + 15y = -70$$

Standard form, in addition to being tidy-looking, is a great form for graphing by intercept-intercept. If $y = 0$, solving $8x = -70$ will give you the x-intercept, and if $x = 0$, solving $15y = -70$ will give you the y-intercept.

 CHECK POINT

Put each equation in standard form.

11. $y = 6x - 7$

12. $y - 7 = 3(x + 2)$

13. $y = -3x + 5$

14. $y + 1 = -4(x - 3)$

15. $4y = 8x - 32$

16. $y - 5 = \frac{1}{2}(x - 12)$

17. $5y = 3 - 4x$

18. $2y + 7 = -6(x + 2)$

19. $\frac{1}{2}y = \frac{3}{4}x - \frac{7}{4}$

20. $5(y - 2) = -3(x - 6)$

Point-Slope Form

The *point-slope form* is a powerful tool when you want to find the equation of a line, but you will rarely be asked to put an equation into this form. If a line has a slope, m, and passes through the point (x_1, y_1), the point-slope form of the line is $y - y_1 = m(x - x_1)$. The equation of a line with a slope of -4 through the point $(5, 2)$ is $y - 2 = -4(x - 5)$.

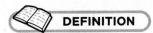

DEFINITION

Point-slope form, $y - y_1 = m(x - x_1)$, is a version of the linear function that defines the relationship between the variables x and y in terms of the slope of the line, m, and one point known to be on the line, (x_1, y_1). If a line with a slope of -3 passes through the point (4, -7), the equation of that line can be written as $y - (-7) = -3(x - 4)$.

CHECK POINT

Identify each equation as slope-intercept form, standard form, or point-slope form.

21. $y - 7 = -2(x + 5)$ 24. $x + y = 10$

22. $3x - 4y = 12$ 25. $y + 2 = \dfrac{3}{5}(x - 4)$

23. $y = 2x - 6$

The slope-intercept form of a linear equation is perfect for graphing by slope and y-intercept. The standard form is useful for graphing by intercepts. The third form, point-slope form, could be used for graphing, but its real power is that it allows you to build the equation of a line from just a little bit of information about its graph.

Finding the Equation

There's more than one way to find the equation of a line. The method you choose depends on what information you have. We'll look at each possible situation.

Given the Slope and *y*-Intercept

If you know that a line has a slope of -3 and its y-intercept is (0, 7), you have very little work to do to find the equation of the line. Just use slope-intercept form, replace m with the slope and b with the y-intercept. $y = mx + b$ becomes $y = -3x + 7$, and you are done.

CHECK POINT

Write the equation of the line with the given slope and y-intercept.

26. slope: -3, y-intercept: 0

27. slope: 5, y-intercept: -9

28. slope: $\frac{1}{2}$, y-intercept: 5

29. slope: $-\frac{4}{3}$, y-intercept: -6

30. slope: 1, y-intercept: 3

Given the Slope and a Point

If you know the slope of the line and a point on the line, but that point is not the y-intercept, you'll want to use point-slope form. This is exactly the situation for which the form is named. If you know that the slope of a line is $\frac{1}{2}$ and it passes through the point (3, 5), you can start with the point-slope form $y - y_1 = m(x - x_1)$ and replace the appropriate pieces with numbers you know. The slope of $\frac{1}{2}$ goes in place of the m, 3 in place of x_1, and 5 in place of y_1, and the form becomes $y - 5 = \frac{1}{2}(x - 3)$.

Technically, you've found the equation of the line at that point, but as mentioned earlier, point-slope isn't really a form you work with. So let's continue on to put this equation into slope-intercept form. (You could do standard form if you prefer.) Start by distributing the $\frac{1}{2}$.

$$y - 5 = \frac{1}{2}(x - 3)$$
$$y - 5 = \frac{1}{2} \cdot x - \frac{1}{2} \cdot \frac{3}{1}$$
$$y - 5 = \frac{1}{2}x - \frac{3}{2}$$

Then add 5 to both sides to isolate the y.

$$y - 5 + 5 = \frac{1}{2}x - \frac{3}{2} + 5$$
$$y = \frac{1}{2}x - \frac{3}{2} + \frac{10}{2}$$
$$y = \frac{1}{2}x + \frac{7}{2}$$

You might wonder why point-slope form is needed. If you can look at the graph to see the slope and a point, why not just look to see what the *y*-intercept is? The previous equation tells you what the problem might be. There's no guarantee that the *y*-intercept will be an integer and if it's not, estimating the fraction can be difficult. Point-slope form lets you use any convenient point.

CHECK POINT

Find the equation of the line with the given slope, passing through the given point. Put the equation in slope-intercept form.

31. Slope: -4, point: (-1, 7)

32. Slope: 5, point: (-2, 6)

33. Slope: $\frac{1}{3}$, point: (9, 9)

34. Slope: $-\frac{5}{4}$, point: (-8, 5)

35. Slope: -1, point: (9, -4)

36. Slope: 2, point: (5, 7)

37. Slope: $\frac{3}{2}$, point: (-8, 3)

38. Slope: 1, point: (-4, -3)

39. Slope: $\frac{2}{5}$, point: (-5, 1)

40. Slope: 3, point: (6, -5)

If you know the slope and the *y*-intercept, use slope-intercept form. If you know a point and the slope, use point-slope form, and simplify. That seems fairly straightforward. But the information you have to work with isn't always as tidy as that.

Suppose you have reason to believe that there is a linear relationship between an adult's height and weight. You think that weight is a linear function of height, but you don't have a rule for that relationship, and no one has conveniently told you the slope of the line. All you know is this particular adult weighed 120 pounds when she was 64 inches tall and 135 pounds when she was 67 inches tall. You have two points: (64, 120) and (67, 135). How do you find the equation of the line?

Given Two Points

If you know two points on a line, you have part of the information you need. You have a point; in fact, you have two. What you need is the slope. You could plot the two points you have on a graph, draw the line, and try to count the boxes to find the rise and run. Sometimes that's a very efficient method, but in a case like this example, when the numbers are large, or if there are lots of fractions or decimals, it could be difficult.

TIP

The slope formula is usually given as $m = \dfrac{y_2 - y_1}{x_2 - x_1}$ but if it's more convenient, you can use $m = \dfrac{y_1 - y_2}{x_1 - x_2}$. It's not important which point goes first, but it is important that you be consistent.

To find the slope of the line efficiently, you can calculate it using a formula. The slope of a line through the points (x_1, y_1) and (x_2, y_2) is $m = \dfrac{y_2 - y_1}{x_2 - x_1}$. The rise of the line can be found by subtracting the y-coordinates and the run by subtracting the x-coordinates. Then the slope is $\dfrac{\text{rise}}{\text{run}}$.

ALGEBRA TRAP

Be careful to keep the same order of subtraction in both the numerator and denominator. It doesn't matter which point you call (x_1, y_1) and which you call (x_2, y_2), but if you switch the order from the numerator to the denominator, the sign of your slope will be wrong.

In the example, the points are (64, 120) and (67, 135). To find the slope of the line, you can use the formula $m = \dfrac{y_2 - y_1}{x_2 - x_1}$ with (x_1, y_1) equal to (64, 120) and (x_2, y_2) equal to (67, 135). So the slope is

$$m = \frac{y_2 - y_1}{x_2 - x_1}$$
$$m = \frac{135 - 120}{67 - 64}$$
$$m = \frac{15}{3}$$
$$m = 5$$

That's saying each time the height increases by 1 inch, the weight increases 5 pounds.

Now you have a slope, $m = 5$, and your choice of either one of your two points, (64, 120) or (67, 135), which means you have enough information to use point-slope form to find the equation of the line. Start with the point-slope form, $y - y_1 = m(x - x_1)$, and substitute 5 for the slope. $y - y_1 = 5(x - x_1)$. Then choose one of the points—either one, but only one. Let's use (64, 120). The 64 goes in place of x_1 and the 120 goes in place of y_1, so the form becomes $y - 120 = 5(x - 64)$. All that's left to do is simplify.

$$y - 120 = 5(x - 64)$$
$$y - 120 = 5x - 5 \cdot 64$$
$$y - 120 = 5x - 320$$
$$y - 120 + 120 = 5x - 320 + 120$$
$$y = 5x - 200$$

You may think using the other point would give you a different equation, but that's not the case. The equation will look different at first. If you used (67, 135) instead of (64, 120), you would start with $y - 135 = 5(x - 67)$ instead of $y - 120 = 5(x - 64)$. That seems different, but when you simplify, it turns out to be the same equation.

$$y - 135 = 5(x - 67)$$
$$y - 135 = 5x - 335$$
$$y = 5x - 200$$

CHECK POINT

Find the slope of the line that passes through the two points given.

41. (6, 3) and (8, -7)   44. (2, -4) and (-4, 8)

42. (-3, 5) and (1, 9)   45. (7, 0) and (0, 7)

43. (-4, -5) and (-1, -2)

Find the equation of the line that passes through the two points given. Put the equation in point-slope form.

46. (5, -10) and (-5, 20)   49. (3, 5) and (-3, 3)

47. (-4, -9) and (6, -4)   50. $\left(-2, \dfrac{21}{2}\right)$ and $\left(1, \dfrac{33}{4}\right)$

48. (4, -5) and (1, 7)

In a perfect world, you know the slope of your line and its y-intercept, drop them into slope-intercept form and you're done. In a little less perfect world, you know the slope and a point, you plug them into point-slope form and simplify. But life is never perfect, so perhaps you only know two points, but you find the slope, plug a point and the slope into point-slope form and simplify. Ready for one last variant?

Parallel and Perpendicular Lines

Can you find the equation of a line parallel to $y = 3x - 7$? Remember, first of all, what parallel means. Two lines in a plane are parallel if they never intersect. They are always the same distance apart. In order for that to happen on a graph, the lines have to cross the plane at the same angle. They must have the same slope.

That's actually helpful. If parallel lines have the same slope, a line parallel to $y = 3x - 7$ also has a slope of 3, but where is this line? You need to know some point to anchor the line. If you have a point for the line to pass through, and you can deduce that the lines have the same slope, a point and a slope are all the information you need.

So what's the equation of a line that passes through (1, 1) and is parallel to $y = 3x - 7$? Parallel lines have the same slope so the slope is 3, and you can use point-slope form.

$$y - y_1 = m(x - x_1)$$
$$y - 1 = 3(x - 1)$$
$$y - 1 = 3x - 3$$
$$y = 3x - 2$$

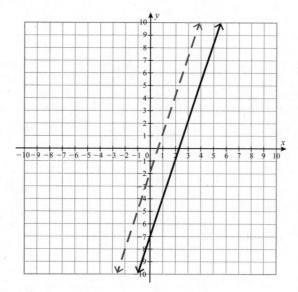

The line $y = 3x - 2$ passes through (1, 1) and is parallel to $y = 3x - 7$.

ALGEBRA TRAP

Parallel lines have the same slope, but before you decide what the slope of the given line is, be sure it's in slope-intercept form. A line parallel to $2x - 3y = 6$ doesn't have a slope of 2, because $2x - 3y = 6$ doesn't have a slope of 2. When you put $2x - 3y = 6$ in slope-intercept form, it becomes $y = \frac{2}{3}x - 2$, and you can see the slope is really $\frac{2}{3}$.

How about a line perpendicular to $y = -2x + 1$? You'll need a point to anchor this one, too. Let's use (1, 1) again, but the slope isn't quite so simple. Perpendicular lines, lines that meet at a right angle, don't have the same slope. If they had the same slope, they'd be parallel and wouldn't cross at all.

To create perpendicular lines, there would have to be one line that rises and one that falls, so the slopes would have to have opposite signs. The rises and runs have to be just right to get exactly a right angle. It turns out that perpendicular lines have slopes that are negative reciprocals. That means their slopes multiply to -1. If the slope of one line is 2, the slope of a line perpendicular to it is $-\frac{1}{2}$. If the slope of one line is $-\frac{5}{3}$, a line perpendicular to it has a slope of $\frac{3}{5}$.

ALGEBRA TRAP

If the slope of one line is 0, it's impossible to find a negative reciprocal for the slope of the perpendicular, because division by 0 is impossible. But if the slope of one line is 0, the line is horizontal, and so a line perpendicular to it is vertical. The equation of a vertical line is $x =$ a constant.

If you want a line that passes through (1, 1) and is perpendicular to $y = -2x + 1$, you can use a slope of $\frac{1}{2}$ and plug the point and the slope into point-slope form.

$$y - y_1 = m(x - x_1)$$
$$y - 1 = \frac{1}{2}(x - 1)$$
$$y - 1 = \frac{1}{2}x - \frac{1}{2}$$
$$y = \frac{1}{2}x + \frac{1}{2}$$

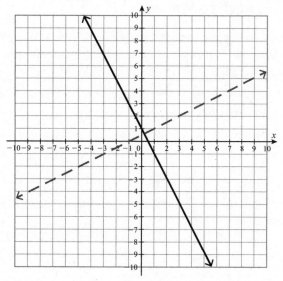

The line $y = \dfrac{1}{2}x + \dfrac{1}{2}$ passes through (1, 1) and is perpendicular to $y = -2x + 1$.

 CHECK POINT

51. Find the equation of a line through the point (6, -3) parallel to $y = 2x - 7$.

52. Find the equation of a line through the point (-3, 2) parallel to $3y = x + 12$.

53. Find the equation of a line through the point (-4, -3) parallel to $y = 4x + 3$.

54. Find the equation of a line through the point $\left(\dfrac{1}{2}, \dfrac{1}{2}\right)$ parallel to $y = -6x + 3$.

55. Find the equation of a line through the point $\left(\dfrac{3}{2}, 4\right)$ parallel to $y = \dfrac{4}{3}x - 1$.

56. Find the equation of a line through the point (4, 7) perpendicular to $y = 2x - 1$.

57. Find the equation of a line through the point (6, -2) perpendicular to $y = -3x + 5$.

58. Find the equation of a line through the point (-1, 7) perpendicular to $x - 2y = -12$.

59. Find the equation of a line through the point $\left(-\dfrac{1}{3}, \dfrac{2}{3}\right)$ perpendicular to $y = \dfrac{1}{3}x + 7$.

60. Find the equation of a line through the point (4, -3) perpendicular to $y = x + 4$.

The Least You Need to Know

- A linear equation can be written in slope-intercept form, $y = mx + b$, where m is the slope and b is the y-intercept, or in standard form, $ax + by = c$, where a, b, and c are integers and a is positive.

- Point-slope form, $y - y_1 = m(x - x_1)$, can be used to create the equation of a line with slope m through point (x_1, y_1).

- If you know two points, (x_1, y_1) and (x_2, y_2), you can find the slope with the formula $m = \dfrac{y_2 - y_1}{x_2 - x_1}$.

- Algebraic techniques can be used to change from one form to another.

- Parallel lines have the same slope. Perpendicular lines have slopes that are negative reciprocals.

Variations on the Line

Those detective stories I love may start out with a straight path to a solution, but of course, that doesn't stay straight or simple for long. Soon things crop up that are a little out of the ordinary, that are exceptions to the rule, or perhaps things just get a little more complicated with another crime or another suspect showing up.

In Part 3, we're going to look at some variations on that linear path we started on. The first bend in our road will be absolute value functions, whose graphs look like lines that took a sharp turn, and whose solutions demand that we consider multiple possibilities. We'll look at vertical and horizontal lines, which bend the rules we learned for lines, and then we'll explore systems of linear equations, which bring together two perfectly reasonable equations and ask us to figure out when they can both be true.

Absolute Value Functions

If it's a little less than 2,800 miles from New York to Los Angeles, how far is it from Los Angeles to New York? It's also a little less than 2,800 miles, of course. It doesn't matter whether you're traveling east to west or west to east; the distance is the same. In algebra, when you think about the real numbers on a number line, direction usually does matter. Moving from -4 to 7 is thought of as a positive move, but 7 to -4 is looked upon as negative. When direction doesn't matter, you're talking about absolute value. In previous chapters, we've talked about the absolute value of a number, but in this chapter, we'll look at what happens when an equation or inequality involves the absolute value of a variable or expression.

In This Chapter

- Solving absolute value equations
- Solving absolute value inequalities
- Solving problems with absolute value equations and inequalities

Understanding Absolute Value

The *absolute value* of a number is its distance from 0 without regard to direction. A number and its opposite have the same absolute value, because they are the same distance from zero, but in different directions.

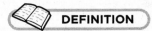

DEFINITION

Absolute value is a measure of distance, without regard to direction. It tells the size, or magnitude, of the number without considering its sign.

When a variable is placed inside absolute value signs, it's more difficult to interpret, because you don't know the value of the variable. You don't know the sign of the number, but rather its distance from zero. If you say $|x| = 4$, you accept both 4 and -4 as possible values of x, because both are four units from zero.

Expressions find their way inside the absolute sign when you think about things like the distance from New York to Los Angeles. East to west, or west to east, it's the absolute value, not the direction, you care about. If x is a number 7 units from 9, then you calculate $9 - x$ or $x - 9$ to get the distance between x and 9. One will give you 7 and one will give you -7. Without knowing the value of x, you can't know which is positive and which is negative, but you don't really care about the direction. What you're saying is $|x - 9| = |9 - x| = 7$. Either subtraction gives you a number that has an absolute value of 7.

When you're trying to translate a statement about distances into an absolute value equation, first find the distance. This is what the absolute value will equal. If Jennie is visiting a town 100 miles from Omaha, Nebraska, the distance is 100 miles, and the equation is going to have the form $|\text{some calculation}| = 100$. If y is a number 8 units from 3, the distance is 8 and the equation will have the form $|\text{some calculation}| = 8$. Next look at the calculation you need to do. Define a variable if it hasn't already been done for you. Usually in these cases the calculation is a subtraction. $|y - 3| = 8$ or $|3 - y| = 8$ will both work.

CHECK POINT

Write an absolute value equation for each sentence.

1. A number n is 6 units away from 9.

2. The number -4 is 10 units away from a number t.

3. The distance between x and 5 is 12 units.

4. Twice the distance between y and -4 is 12.

5. The distance between z and 12 is 4 more than z.

Write a sentence that describes the relationship in the equation.

6. $|x - 7| = 3$

7. $|t - (-2)| = 9$

8. $|8 - y| = 11$

9. $5|x - 4| = 20$

10. $|z - 7| = z - 3$

Understanding absolute value and being able to write absolute value equations are important steps, but equations aren't especially helpful unless you can solve them. Solving absolute value equations is a lot like solving linear equations, but not quite the same. Let's have a look at the similarities and differences.

Solving Absolute Value Equations

We talk a lot about the absolute value of a number when we learn to work with integers, but it's also possible to have a variable expression inside an absolute value sign. If you write $|x| = 4$, that's not quite the same as saying $x = 4$. It's true that replacing x with 4 makes the statement true. The number 4 has an absolute value of 4, but it's also true that the number -4 has an absolute value of 4. The equation $|x| = 4$ has two solutions, $x = 4$ and $x = -4$.

Most equations that have a variable expression inside absolute value signs will have two solutions, so when you encounter one of these absolute value equations, expect that. Look for the two solutions by translating the absolute value equation into two equations that do not involve absolute value. Take the expression inside the absolute value signs and show that it could have a positive value or a negative value. The absolute value equation $|x| = 4$ becomes $x = 4$ or $x = -4$. The absolute value equation $|4x - 3| = 9$ gets rewritten as $4x - 3 = 9$ or $4x - 3 = -9$.

Once you have rewritten the absolute value equation as two separate equations, you can solve them as you normally would.

$$|4x-3|=9$$

$4x-3=9$	$4x-3=-9$
$4x=12$	$4x=-6$
$x=3$	$x=-\dfrac{3}{2}$

This gives two possible solutions.

To solve $|3x-7|=5$, treat it as though it were two equations, one that says $3x-7=5$ and one that says $3x-7=-5$. Solve both equations.

$$3x-7=5 \quad \text{and} \quad 3x-7=-5$$
$$3x=12 \qquad \qquad 3x=2$$
$$x=4 \qquad \qquad x=\dfrac{2}{3}$$

ALGEBRA TRAP

Not every absolute value equation has opposites as its solutions. Write the two equations that are equivalent to the absolute value statement, and don't just solve one and assume the other will give you the opposite. Take the time to solve them both, independently.

CHECK POINT

Solve each absolute value equation. Be sure to check your answers in the original equation.

11. $|3x-7|=2$

12. $|2t+5|=3$

13. $|y-8|=1$

14. $|x-13|=5$

15. $|11+8x|=-17$

16. $2|3x-1|=14$

17. $|2y-4|+7=9$

18. $5-3|x+1|=-4$

19. $2+7|4-3x|=16$

20. $|4x-5|=x$

Solving Absolute Value Inequalities

Like absolute value equations, inequalities involving absolute values need special treatment. If you write $|x| < 4$, it's not the same as writing $x < 4$. On the positive side, x might certainly be any number between 0 and 4, but what about negative numbers? Can x be replaced with any negative number to get a true statement? It doesn't take much trying to find out the answer is no. Numbers between 0 and -4 will work, but numbers less than or equal to -4 will make the statement false. The absolute value of -5 is not less than 4. The inequality $|x| < 4$ is equivalent to the compound inequality $-4 < x < 4$. These are the values of x that are less than 4 units from 0.

If you change the direction of the inequality, however, so that you have $|x| > 4$, the values of x that make true statements are those greater than 4 or less than -4. Once again, the absolute value inequality becomes a compound inequality, but this time, it's an *or* inequality, $x < -4$ or $x > 4$, rather than an *and* inequality.

To remember how to change an absolute value inequality to the correct compound inequality, use the mnemonic device GreatOR/Less ThAND. (It's a deliberate misspelling.) If the absolute value is *greater* than a number, make an *or* inequality. If the absolute value is *less than* a number, make an *and* inequality.

To solve an absolute value inequality, translate the absolute value inequality to a compound inequality. If the absolute value is greater than a constant, use an *or* compound. If the absolute value is less than a constant, use an *and* compound.

GreatOR:	|expression| > constant	expression > constant or expression < −constant
Example:	$|3x - 7| > 5$	$3x - 7 > 5$ or $3x - 7 < -5$
Less ThAND:	|expression| < constant	−constant < expression < constant
Example:	$|4x + 5| < 17$	$-17 < 4x + 5 < 17$

CHECK POINT

Solve each absolute value inequality. Graph the solution set on the number line.

21. $|6x - 7| > 5$

22. $|1 - 6x| < 7$

23. $|5y + 2| \le 12$

24. $|7z - 11| \ge 3$

25. $|8a + 3| \le 19$

26. $|5 - 4x| < 21$

27. $|16p + 22| \ge 6$

28. $|9t + 5| \le -2$

29. $|2x + 10| < 32$

30. $|7y - 23| \ge 0$

Modeling with Absolute Value

When do you need absolute value to solve a problem? Your first look at the information in a problem may suggest an equation to solve the problem, but a little more thought may tell you that there is more than one possible answer, and that's the clue that you may need an absolute value equation or inequality.

Absolute value problems are usually about distance in some way. It may be the actual distance between two places, or the distance between points on a number line, or an amount of a change. Whatever the context, the direction isn't known, or isn't significant, so absolute value is appropriate.

When the two points in the coordinate plane have the same x-coordinate, they sit on the same vertical line. If two points have the same y-coordinate, the points sit on a horizontal line. The points (3, 0) and (-12, 0) both sit on the horizontal line called the x-axis, and you can think about them as though they were points on the number line. The points (4, -7) and (4, 9) both lie on the vertical line $x = 4$, and you can find the distance by focusing on the y-coordinates.

Suppose you are looking for all the points on the x-axis that are exactly 3 units from the point (5, 0). You're looking for points on the x-axis, so you know the points will have a y-coordinate of 0. Let the points you're looking for be represented by $(x, 0)$. The distance between (5, 0) and $(x, 0)$ is 3, but you don't know if $(x, 0)$ is to the right of (5, 0) or to the left. You can write an absolute value equation about the x-coordinates.

$$|5 - x| = 3$$

Solving that equation will give you two x-coordinates.

$$|5 - x| = 3$$

$$
\begin{array}{ll}
5 - x = -3 & \qquad 5 - x = 3 \\
-x = -8 & \qquad -x = -2 \\
x = 8 & \qquad x = 2
\end{array}
$$

Each of these x-coordinates gives you a point that is 3 units from (5, 0). The points are (8, 0), three units to the right of (5, 0), and (2, 0), three units to the left of (5, 0).

CHECK POINT

Use an absolute value equation or inequality to solve each problem.

31. The difference between a number x and 14 is 22. Find all possible values for the number.

32. A number y is 24 units away from -17. Find all possible values of y.

33. If n represents all integers within 3 units of 9, find all possible values of n.

34. The difference between Peter's salary and Paul's is less than $4,000 a year. If Peter earns $82,500 per year, describe the possibilities for Paul's salary.

35. Find all the points on the y-axis that are exactly 6 units from the point (0, -4).

36. Find all the points on the x-axis that are exactly 13 units from the point (-12, 0).

37. Find all the points on the y-axis that are exactly 10 units from the point (0, 3).

38. A body temperature that differs from the average of 98.6°F by more than 2° requires medical attention. What temperatures should receive medical attention?

39. The electricity supplied to most homes in the United States is designed to be 115 volts, but fluctuation within certain ranges is normal and natural. Too great a deviation can cause problems, however. Problems arise when the difference between the actual voltage V and the desired 115 volts exceeds 5 volts. Find the range of acceptable voltages.

40. Polling is used to predict the results of elections based on the responses of a relatively small group of respondents. In addition to giving the results of the poll, they often give a margin of error, which tells how far above or below their result they expect the actual result might be. If a candidate is predicted to earn 46% of the vote with a margin of error of 4%, in what range would you expect the candidate's actual vote total to fall?

Linear equations and linear inequalities describe most of the situations you'll encounter when solving problems, so being able to graph them efficiently is an important skill. If other, unusual relationships arise, we can always rely on the fundamental table of values to graph a new type of equation or inequality.

You know that absolute value equations in one variable can be translated into two linear equations and solved in ways very similar to solving linear equations. Absolute value inequalities translate into compound linear inequalities, so much of the work there is similar. It's not surprising, therefore, that graphs of absolute value functions and absolute value inequalities are produced by methods similar to linear equations and inequalities.

Graphing Absolute Value Equations

Just as solving an absolute value equation involved solving two linear equations, graphing an absolute value equation turns out to mean graphing two linear equations, or at least parts of each of two linear equations. Let's take a first look by means of a table of values.

 TIP

> The graph of an absolute value function is a V-shaped graph made up of two rays that meet at their endpoints. Remember that lines go on forever in both directions. Rays have one endpoint and continue endlessly in one direction. The sides of the V are rays with the same endpoint, which are called the vertex.

To graph $y = |x|$, first you need to choose values for x, including both positive and negative values, and evaluate for y.

x	-3	-2	-1	0	1	2	3
$y = |x|$	3	2	1	0	1	2	3

Plot the points and connect them, but be careful. Don't make assumptions about what the graph should look like. Let the points lead you. Here's what it looks like after plotting the points.

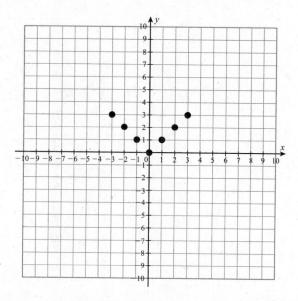

Can you see the two rays? The first few points from the table suggest a falling line, a line with a negative slope, and the last few create a rising line, with a positive slope, but you only want part of each one. You don't want any points with negative y-coordinates. The two rays both have a y-intercept of 0, so they will intersect there at (0, 0). That meeting point is called the *vertex*.

 **DEFINITION**

The **vertex** of an absolute value graph is the point where the two rays meet to form the point of the V-shape.

Don't just grab a straight edge and draw lines from edge to edge of the grid, though. Remember that absolute value is never negative. The graph of $y = |x|$ will never have any points with negative y-coordinates. So connect those falling points, and extend that ray upward, but don't extend it beyond the origin. If you do, you'll be going into negative y-values, and you don't want that. In the same way, the rising portion of the graph can't go below the x-axis. Here's what the graph of $y = |x|$ looks like.

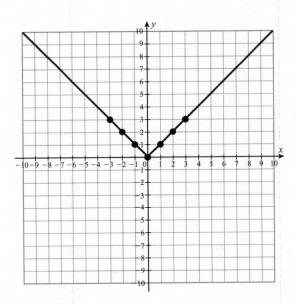

The V shape you see is characteristic of the graph of the absolute value function. Changes to the function's rule may move the V left or right, up or down. The V may be narrow or wide, upright or upside down, but if it's an absolute value function, its graph will have a V shape.

 THINK ABOUT IT

Extending the sides of the absolute value graph in both directions will not only include points you don't really want to include, but will also create a graph that is not a function. The absolute value function assigns to each input only one output: the magnitude of the number without its sign. If you extend the sides through the vertex, you'll be including a second point for each x-coordinate, forming an X-shape that fails the vertical line test.

Any time you face a new type of function, you can always return to a table of values to produce the graph, but of course, it's nice to use a quicker method if you can. The two sides of the V in the graph of the absolute value function are portions of the line $y = x$ and $y = -x$. You know how to graph both of those quickly, so if you had a way to look at an absolute value function and predict what the two lines should be, graphing would be simple.

To find the basic shortcut, you can use what you know about solving absolute value equations. Suppose you wanted to solve $|2x - 7| = 3$. You would write two linear equations and solve each one.

$$|2x - 7| = 3$$

$$
\begin{array}{ll}
2x - 7 = 3 & 2x - 7 = -3 \\
2x = 10 & 2x = 4 \\
x = 5 & x = 2
\end{array}
$$

Now think about an absolute value function like $y = |2x - 7|$. If you rewrite that as $|2x - 7| = y$, you can see that the y is in the place of the 3 in the solution example. So let's start thinking about the function to graph in the same way as the equation to be solved. Translate it into two equations.

$$|2x - 7| = y$$
$$2x - 7 = y \qquad\qquad 2x - 7 = -y$$

One of those equations is just $y = 2x - 7$, but the other needs a bit of cleaning up. You don't want to know what $-y$ equals. You want to know what y equals, so you need to multiply both sides by -1.

$$2x - 7 = -y$$
$$(-1)(2x - 7) = (-y)(-1)$$
$$-2x + 7 = y$$

One side of the V-shaped graph is part of the line $y = 2x - 7$ and the other side is part of the line $y = -2x + 7$.

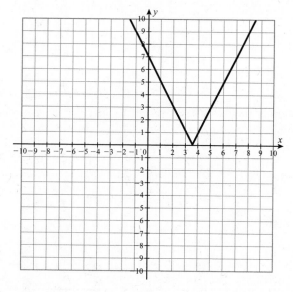

If you need to graph an absolute value function of the form $y = |mx + b|$, graph $y = mx + b$ and $y = -1(mx + b)$. Do not extend either one below the x-axis. The two lines should meet at the x-axis.

That's a basic shortcut, but there is a complication. What if some operations are outside the absolute value signs? How do you graph $y = |2x| - 7$, if part of the expression, the $2x$, is inside the absolute value sign and the -7 is outside? You don't need to start all over. You just need to be clever. First add 7 to both sides, so that the absolute value is on one side and everything that's not in the absolute value is on the other side.

$$y = |2x| - 7$$
$$y + 7 = |2x|$$
$$|2x| = y + 7$$

Now you can tackle it the way you dealt with solving absolute value equations. Rewrite it as two linear equations.

$$|2x| = y + 7$$
$$2x = y + 7 \qquad\qquad 2x = -1(y + 7)$$

The first of those quickly simplifies to $y = 2x - 7$. The second takes a little work.

$$2x = -1(y + 7)$$
$$\frac{2x}{-1} = \frac{-1(y + 7)}{-1}$$
$$-2x = y + 7$$
$$-2x - 7 = y$$

The second equation simplifies to $y = -2x - 7$, and the first to $y = 2x - 7$. You have the positive and the negative version of the expression in the absolute value signs, but the constant term is unchanged. Your graph looks like this.

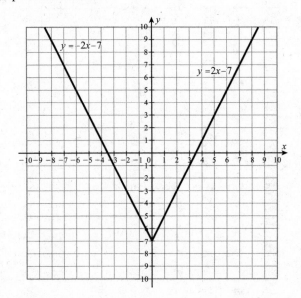

You can create a variant of our shortcut. To graph an absolute value function of the form $y = |mx| + b$, graph $y = mx + b$ and $y = -mx + b$, to form the V shape.

Do you notice that the graph goes below the x-axis, into territory where the y-coordinates are negative? If the entire expression $2x - 7$ were inside the absolute value signs, that shouldn't happen, but because the equation $y = |2x| - 7$ tells you to subtract 7 after finding the absolute value, it is possible to produce negative y-values. Notice how far down the y-intercept is? It's at -7. The absolute value part of the function would have values greater than or equal to zero, but that subtraction afterward pulls the graph down 7 units.

There are a few things you can say about the graph of an absolute value equation just by looking at the equation.

What an Equation Can Tell You About a Graph

Detail of Equation	Example	Effect on the Graph	
$y = |ax|$ (or $y = a|x|$ if $a > 0$)	$y = |4x|$ or $y = 4|x|$	Slope of one side of the V is a, the other is $-1 \cdot a$. The V is in the upright position.	
$y = -a|x|$	$y = -3|x|$	Slope of one side of the V is a, the other is $-1 \cdot a$. The V is inverted.	
$y = |x - b|$	$y = |x - 1|$	The V shifts b units to the right.	
$y = |x + b|$	$y = |x + 5|$	The V shifts b units to the left.	
$y = |x| - c$	$y = |x| - 2$	The V shifts c units down.	
$y = |x| + c$	$y = |x| + 3$	The V shifts c units up.	

CHECK POINT

Graph each absolute value equation.

41. $y = |x - 3|$

42. $y = |x| + 1$

43. $y = |x + 1|$

44. $y = |x| - 5$

45. $y = |2x|$

46. $y = -3|x|$

47. $y = |2x - 5|$

48. $y = |7 - 4x|$

49. $y = |3x - 8|$

50. $y = |12 - 5x|$

The Least You Need to Know

- Absolute value equations have two solutions, because the expression in the absolute value could be positive or negative.

- Absolute value inequalities translate to compound inequalities. Absolute value greater than a constant becomes an *or* inequality, and absolute value less than a constant becomes an *and* inequality.

- Absolute value equations and inequalities are useful in situations where more than one answer is possible.

Special Graphs

In this chapter, we'll investigate vertical and horizontal lines, both of which deviate from the pattern we've been focusing on. We'll also look at some related graphs. We'll look at the graphs of linear inequalities, which are similar to but not quite the same as linear equations. Although they are not technically linear equations, absolute value equations are related, both in terms of how we solve them and in the appearance of their graphs. We'll take a look at the graphs of absolute value equations and inequalities as well.

Vertical and Horizontal Lines

We've talked about linear functions as relationships that fit the pattern $y = mx + b$, a variable term and a constant term. We looked at what happened when b, the y-intercept, was equal to 0. When that happened, the line passed right through the origin. One of the things we haven't looked at is what happens when m, the slope, is equal to 0.

Horizontal Lines

Remember that slope is a measurement of the number of steps a line rises or falls each time it moves one step to the right. Slope is sometimes defined as $\text{slope} = \dfrac{\text{rise}}{\text{run}}$, so a slope of 0 would mean the line doesn't rise at all, no matter how far it runs. That would mean it's a horizontal line. Points on a horizontal line all have the same y-coordinate. The points (-4, 5) and (6, 5) both lie on the horizontal line $y = 5$. You can plug those points into the slope formula.

$$m = \frac{y_2 - y_1}{x_2 - x_1}$$

$$m = \frac{5 - 5}{-4 - 6}$$

$$m = \frac{0}{-10}$$

$$m = 0$$

Because $m = 0$, the equation of the line becomes $y = 0x + b$ or $y = b$. The equation of a horizontal line with a y-intercept of 5 is $y = 5$. The equation of a horizontal line with a y-intercept of -3 is $y = -3$.

ALGEBRA TRAP

Don't confuse dividing zero by a non-zero number with dividing by zero. For any non-zero number c, $\dfrac{0}{c} = 0$, but $\dfrac{c}{0}$ is undefined. Zero divided by a non-zero number is zero, but division by zero is impossible.

If you look at the graph of a horizontal line like $y = 5$, you can see that simple equation actually describes the line very well.

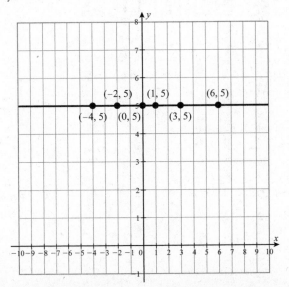

Every point on the line $y = 5$ has a y-coordinate of 5. Every point on the horizontal line $y = -3$ has a y-coordinate of -3. That's enough information to graph the horizontal line.

Vertical Lines

If the equation of a horizontal line is always $y = $ a constant, what does the equation of a vertical line look like? Vertical lines do rise, or fall, depending on your point of view. It's difficult to say which because they don't run. Their x-coordinates never change. Because all the points on a vertical line have the same x-coordinate, vertical lines have the form $x = $ a constant, and the constant is the x-coordinate of each point on the line.

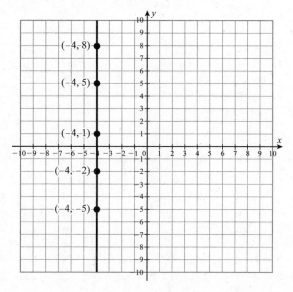

Because the vertical line rises but never runs, something unusual happens when you look at its slope. Slope $= \dfrac{\text{rise}}{\text{run}}$ but for a vertical line, the run is zero, and division by zero is impossible. Using the point (-4, -5) and (-4, 8), you can plug into the slope formula, but you bump into a problem.

$$m = \frac{y_2 - y_1}{x_2 - x_1}$$

$$m = \frac{-5 - 8}{-4 - -4}$$

$$m = \frac{-13}{0}$$

Division by zero is impossible. The expression $\dfrac{-13}{0}$ is undefined. A vertical line has an undefined slope, but people will often say it has no slope, meaning the same thing.

CHECK POINT

Identify each equation as the equation of a vertical line, a horizontal line, or an oblique (slanted) line.

1. $y = 12$ 4. $x = 7$

2. $x = 5y$ 5. $x - 6 = 0$

3. $y = 12 + x$

Graph each line.

6. $x = -4$ 9. $y + 7 = 4$

7. $y = -3$ 10. $x = 1$

8. $x + 4 = 6$

After we talked about solving linear equations, we looked at solving linear inequalities and saw that it was very much the same process, with one key addition. When we multiplied or divided both sides by a negative number, the direction of the inequality sign reversed. We needed that additional exploration because the situations we investigate don't always tell us exactly what things equal. Often real problems talk about "at least" this much or "not more than" that much.

Now that you know how to graph the equation of a linear function, it makes sense to think about what happens when inequalities creep into that graphing process. If you design some t-shirts for your softball team, and order them from a custom printing service, you probably pay a one-time start-up fee plus a cost for each shirt. If the start-up fee is $5 and each shirt is $20, the cost of your order of x shirts would be $y = 20x + 5$. Generally, however, such services will discount the price for larger orders. If you have enough people on your team, that $20 per shirt price may be reduced, so the total cost of your order is better represented as $y \leq 20x + 5$. How do you graph that?

Graphing Linear Inequalities

Just as solving inequalities is very similar to solving equations, graphing inequalities has a lot in common with graphing equations. The graph of an equation like $y = 20x + 5$ is a picture of all the points whose y-coordinates are exactly equal to $20x + 5$. If you want to graph $y \leq 20x + 5$, you want those points, but you also want the points whose y-coordinates are smaller than $20x + 5$. You could start our graph by drawing the graph of the equation $y = 20x + 5$, but then you'd also need to show those other points.

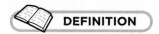 **DEFINITION**

A **linear inequality** is a statement that defines a relation (not a function) in which the output is less than, or greater than, or less than or equal to, or greater than or equal to, some expression involving the input value.

When you create the graph of a *linear inequality*, you actually divide the coordinate plane into three sections: the points on the line, the points above the line, and the points below the line. The points on the line $y = 20x + 5$ are the points whose y-coordinate is exactly $20x + 5$. The points above the line have larger y-values, so they are the points where $y > 20x + 5$, and the points below the line are the points whose y-coordinates are smaller than $20x + 5$, so they are the points where $y < 20x + 5$.

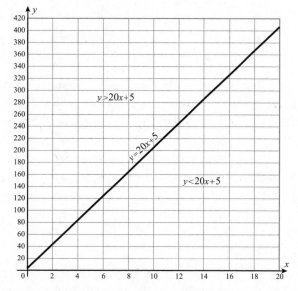

Graphing a linear inequality is similar to graphing linear equations but with two additional considerations. You start by graphing the line as though you were graphing an equation. The first additional element comes up as you go to draw the line. If the inequality is a "less than or equal to" inequality or a "greater than or equal to" inequality, your solution should include the points of the line, so draw a solid line. If the inequality doesn't include the "or equal to" part, you don't want to include the line in your solution, but you need to know where it is, because it serves as a boundary between the points you want and the points you don't want. So go ahead and draw it in, but use a dotted line.

To draw the graph of $y > 2x - 7$, start from the y-intercept of -7, count out the slope of up 2, 1 right, a few times, placing dots. You'll shade some points in a moment, but because you have $y > 2x - 7$, not $y \geq 2x - 7$, you don't want to include the points on the line. Draw a dotted line, not a solid line, to signal that you don't mean to include this line.

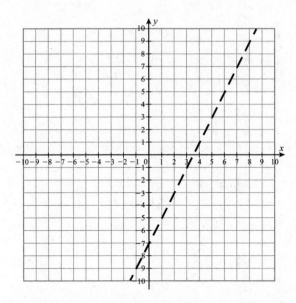

Once you know where to find the points for which the *y*-coordinate is exactly 7 less than twice the *x*-coordinate, you need to find and shade the points for which the *y*-coordinate is more than $2x - 7$. When the inequality is in slope-intercept form, with the *y* isolated, you can say that the points where *y* is greater are above the line. If you can't put the inequality in that form, don't guess at where to shade. Pick a point—the origin is often an easy one to choose, as long as it is not on your line—and plug the *x*- and *y*- coordinates of that point into the inequality.

$$y > 2x - 7$$
$$0 > 2 \cdot 0 - 7$$
$$0 > -7$$

If the result you get is true, the point you chose is in the region you need to shade. Substituting 0 for *x* and 0 for *y* gave you a true statement, so the origin (0, 0) is in the shaded region. Shade that side of the line. If the statement had come out false, you would shade the opposite side of the line.

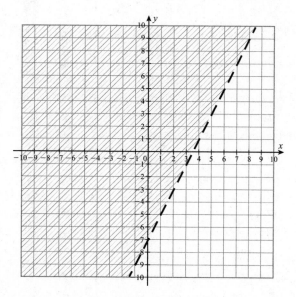

Follow these steps to graph a linear inequality.

1. Isolate y.

2. Graph the line. Draw the line very lightly.

3. If the inequality is an "or equal to" make the line a solid line. If not, use a dotted line.

4. If the inequality begins with $y \geq$ or $y >$, you will shade all the points above the line. If it begins with $y \leq$ or $y <$, you will shade all the points below the line.

5. To be certain that you're shading the correct area, choose a point from the region you think should be shaded, and substitute its coordinates for x and y in the original inequality. If the result is a true statement, that is the correct region to shade.

 CHECK POINT

Graph each inequality.

11. $y \leq 2x - 1$

12. $y > \dfrac{1}{2}x + 3$

13. $y < -3x + 5$

14. $y \geq -\dfrac{2}{3}x + 6$

15. $x < 3$

16. $y - 5 \geq 0$

17. $x + y \leq 8$

18. $2x - y > -6$

19. $4x - 3y < 12$

20. $x \leq 2y + 8$

Graphing Absolute Value Inequalities

Just as you can graph a linear inequality by adding a few steps to the process for graphing a linear equation, you can graph an absolute value inequality with just a few more steps than it takes to graph an absolute value equation. Whether you choose to graph an absolute value equation by a table of values or you use some shortcuts to produce the graph, creating that V-shaped graph is the place to start when you're called upon to graph an absolute value inequality.

Once you have the basic graph in place, the remaining steps are the same ones you would take with linear inequalities. First, you must examine the inequality sign to see if it includes an "or equal to." If your inequality sign is ≥ or ≤, you'll draw the V-shaped absolute value graph with a solid line. If the inequality sign is > or <, you'll use a broken or dotted line to show that you don't want to include the points on the V.

The last consideration is where to shade, and this is when graphing absolute value inequalities gets a little bit tougher than graphing linear inequalities. The rule is the same. If you want the points whose y-coordinates are smaller, shade down. If you want the points with larger y-coordinates, shade up. The shape of the absolute value graph, however, can make the choice confusing at times.

 THINK ABOUT IT

A line drawn in the coordinate plane creates two half-planes, but that name is somewhat misleading. Each region goes on forever. There's no edge to the paper, or corner to page to define the boundaries of the sections. You really can't take half of forever.

A line clearly divides the coordinate plane. There are points above the line, points on the line, and points below the line. The same is actually true of an absolute value graph. There are points on the V, points above the V and points below the V, but we don't really think in terms of above and below a V. We might think about inside and outside the V, as if it were a cup.

The problem with that image is that if the V is in its upright position, the points with greater y-coordinates are inside the V, but if the V is inverted, the points with smaller y-coordinates are inside. That could get too confusing.

The better plan is to plot the absolute value equation, using a solid or dotted line as appropriate. If the inequality says that y is greater than the expression, place your pencil on the V and pull the shading up. If y is less than the absolute value expression, place your pencil on the V and pull down.

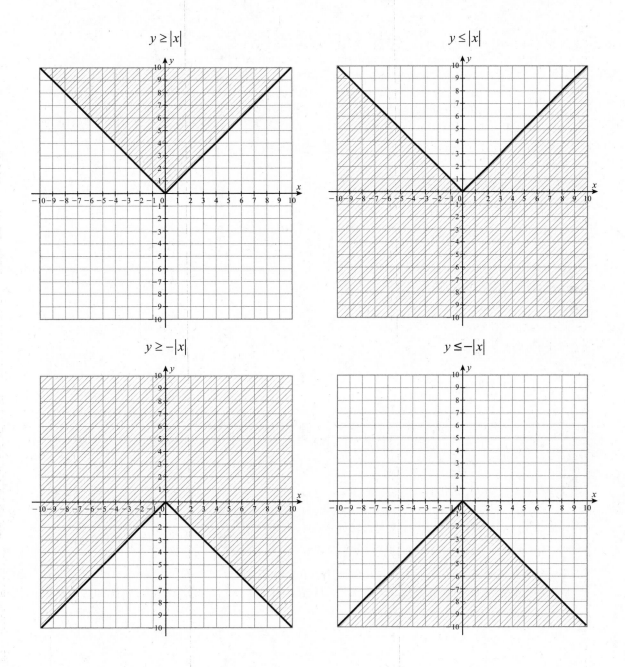

$y \geq |x|$

$y \leq |x|$

$y \geq -|x|$

$y \leq -|x|$

Let's look at the process by graphing the absolute value inequality $y < 4 - |x + 3|$.

1. Start by putting the inequality in the most convenient form. $y < 4 - |x + 3|$ will be easier to graph if you write it as $y < -|x + 3| + 4$.

2. Inspect the inequality for clues to the graph. The negative sign in front of the absolute value sign tells you the graph will open down, and the absence of any coefficient on x or multiplier in front says the slopes will be 1 and −1. The $x + 3$ tells you the graph will move 3 units left and the 4 added on the end moves the graph up 4 units.

3. Make a table of values if you don't feel you have enough information.

4. Sketch the graph.

5. Check the inequality sign. There is no "or equal to" in $y < -|x + 3| + 4$, so use a dotted line to draw the V.

6. Check the direction of the inequality. This one is "less than," so you will shade down.

Here's your graph.

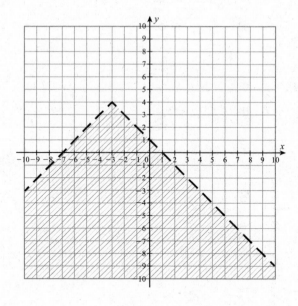

CHECK POINT

Graph each absolute value inequality.

21. $y < |x - 7|$

22. $y \geq |x + 3|$

23. $y > |x| - 4$

24. $y \leq |x| + 1$

25. $y > \left|\dfrac{3}{2}x\right|$

26. $y \geq |2x| - 3$

27. $y < \left|\dfrac{1}{2}x - 3\right|$

28. $y \leq \left|\dfrac{1}{2}x\right| - 3$

29. $y \geq -2|x| + 5$

30. $y > -3|x + 4| + 5$

The Least You Need to Know

- Horizontal lines have a slope of 0. The equation of a horizontal line is $y = $ a constant.

- Vertical lines have an undefined slope. The equation of a vertical line is $x = $ a constant.

- To graph a linear inequality, graph the line using a solid line if the inequality is ≤ or ≥, and a dotted line if the sign is < or >. Shade up if y is greater than the expression. Shade down if y is less than the expression.

- To graph an absolute value inequality, graph the absolute value equation using a solid line if the inequality is ≤ or ≥, and a dotted line if the sign is < or >. Shade up if y is greater than the expression. Shade down if y is less than the expression.

Systems of Equations and Inequalities

Imagine you're standing in line to buy lunch. You want a hot dog and a soda, but you don't see the prices listed anywhere. There are two people in front of you. The first one orders 3 hot dogs and 2 sodas and is charged $8. The person just in front of you orders 4 hot dogs and 2 sodas and pays $10. How much will you pay for your hot dog and soda?

Until now, we've talked about solving equations with one variable and graphing equations with two variables. That general rule holds because there's no way to be certain of a solution when you have one equation with two variables. If the only information you had was the order of the first person in line, 3 hot dogs and 2 sodas for $8, or $3x + 2y = 8$, you could find more than one set of values for x and y that might work. You'd have no way of knowing which combination is correct. But when you add that second order to your information, you have a chance to come up with one right answer.

In this chapter, we'll look at algebraic techniques for finding that one right answer. Guess and test can sometimes be helpful, but it's time consuming, so we'll look at three strategies for solving problems in which you're looking for

In This Chapter

- Solving systems of linear equations by graphing
- Solving systems of linear equations by substitution and linear combinations
- Solving problems using linear systems
- Solving systems of inequalities by graphing

two variables and you have two pieces of information, or two equations, to work with. The first is a natural for two variable equations, because it's a graphing method. The other two methods are based on the techniques you use to solve one equation.

Solving Linear Systems

Two linear equations with the same two variables form a *linear system* of equations. The key to solving for the variables is to use both equations at the same time, which is why systems are sometimes called simultaneous equations. There are three methods for finding the solution of a system: graphing, substitution, and combination.

 **DEFINITION**

A **linear system** is a set of two linear equations, each involving the same two variables. The solution of the system is the one pair of values that satisfy both equations.

Solving by Graphing

The graph of an equation is a picture of all the points that represent pairs of numbers that balance, or solve, the equation. If you graph two equations on the same set of axes, either the lines will be parallel, or they will turn out to be the same line, or they will cross. If they're parallel, there is no point on both lines, so there's no pair of numbers that solve both equations. The system has no solution if the lines are parallel. If they are the same line, they have infinitely many points in common so the system has infinitely many solutions. If the lines intersect, that one point where they cross represents the one pair of numbers that satisfies both equations. The one point common to both lines is the solution to the system.

Remember the situation in the hot dog line? You can write an equation to describe the transaction of each of the customers ahead of you. Let x stand for the number of hot dogs a customer buys and let y stand for the number of sodas. The customer who orders 3 hot dogs and 2 sodas and is charged $8 gives you the equation $3x + 2y = 8$. The person who orders 4 hot dogs and 2 sodas and pays $10 gives you the equation $4x + 2y = 10$. To figure out what your one hot dog and one soda will cost, you need to solve the system

$$3x + 2y = 8$$
$$4x + 2y = 10$$

To find the solution by graphing, first put each equation in slope-intercept form.

$$3x + 2y = 8 \quad \Rightarrow \quad y = -\frac{3}{2}x + 4$$
$$4x + 2y = 10 \quad \Rightarrow \quad y = -2x + 5$$

Graph each equation. The first has a y-intercept of 4 and a slope of $-\dfrac{3}{2}$. The second equation has a y-intercept of 5 and a slope of -2. Focus on the first quadrant for this problem, because the number of hot dogs and the number of sodas cannot be negative numbers.

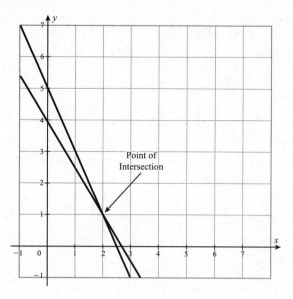

The point at which the lines cross represents the solution of the systems. In this case, the point of intersection is (2, 1), which means $x = 2$ and $y = 1$ is the solution of the system. That means hot dogs cost $2 and sodas cost $1. If you buy one of each, you'll pay $3.

 CHECK POINT

Solve each system of equations by graphing both equations and locating the point of intersection. Check your answers in both equations. If the lines are parallel, say that the system has no solution.

1. $y = 2x + 3$
   $y = 3x - 1$

2. $y = -\dfrac{1}{2}x + 4$
   $y = \dfrac{3}{4}x - 1$

3. $x + y = 9$
   $x - y = 1$

4. $3x - y = 5$
   $6x - 2y = 8$

5. $2x - 5y = -44$
   $y = 6 - x$

6. $y = 3x - 7$
   $2x - 3y = 14$

7. $y = 3x - 5$

    $y = -\dfrac{1}{3}x + 5$

8. $2x + y = -8$

    $3x - 5y = 14$

9. $x + y = 6$

    $y = -x - 4$

10. $3x - 2y = 5$

    $x + 4y = -3$

The graphing method can often be quick and simple, but it does have its drawbacks. If the solution involves extremely large or extremely small numbers, it's hard to find a way to scale the axes that will let you see the intersection on the page and still have enough detail to see exactly what the point of intersection is. If your solution is represented by the point (4672, 5928), graphing may not be your best method.

Graphing is also not ideal for solutions that are not whole numbers. If your solution is (3.082, -7.121), looking at a graph will probably make you think it's (3, -7), which might be good enough in some situations, but it's important to have other methods available.

 TIP

Don't be afraid to change the scale on the axes when you draw a graph. One box doesn't have to be 1 unit. You can count by 2s or 5s or 10s or whatever number is convenient if your problem involves large numbers, and you can make each box worth a half or a tenth, if your numbers are small. There's no rule that says you have to use the same scale on the vertical and horizontal axes, but if they're different, you need to be very careful when counting slope.

Solving by Substitution

You would not be able to find a unique solution to a single equation with two variables, like $3x - 4y = 23$, but if I tell you that $y = -5$, you can replace y with -5, and then you'll be able to solve.

$$3x - 4y = 23$$
$$3x - 4 \cdot (-5) = 23$$
$$3x + 20 = 23$$
$$3x = 3$$
$$x = 1$$

The substitution method for solving a system works on a similar idea. In this method, you use one equation to tell you what one of the variables equals, and then you substitute that into the other equation.

$$2x - 5y = -44$$
$$y = 6 - x$$

To solve this system by the substitution method, you can use the bottom equation to tell you what y equals. It equals $6 - x$. Substitute $6 - x$ for y in the top equation.

$$2x - 5y = -44$$
$$2x - 5(6 - x) = -44$$
$$2x - 30 + 5x = -44$$
$$7x - 30 = -44$$

Making that substitution and simplifying gives you an equation you can solve easily.

$$7x - 30 = -44$$
$$7x = -14$$
$$x = -2$$

Now you know that $x = -2$, but you're not done. You still need to find out the value of y, so substitute once more. This time, choose either one of the original equations and substitute -2 for x. Let's use $y = 6 - x$ and replace x with -2.

$$y = 6 - x$$
$$y = 6 - (-2)$$
$$y = 6 + 2$$
$$y = 8$$

Now we have a solution of the system:

$$x = -2 \text{ and } y = 8$$

That example conveniently had one equation that directly told you what y equaled. That's great, but it doesn't always happen. What should you do with a system like this one?

$$3x - 2y = 5$$
$$x + 4y = -3$$

Carefully look over both equations. Your job will be to pick one equation, either one, and isolate one variable, either variable, in that equation. Look for the easiest one. I think using the bottom equation and isolating x may be the way to go. You might very well choose a different equation and/or a different variable and that's fine. It's your choice.

TIP

Some people think you can only isolate y, but that's not the case. It's a mistaken impression that comes from the fact that you often see linear equations in $y = mx + b$ form. If it's easier to isolate x, that's fine.

Taking the bottom equation, you can isolate x by just subtracting $4y$ from both sides.

$$x + 4y = -3$$
$$x = -3 - 4y$$

Now you'll go to the top equation and replace x with $-3 - 4y$. Then you can solve for y.

$$3x - 2y = 5$$
$$3(-3 - 4y) - 2y = 5$$
$$-9 - 12y - 2y = 5$$
$$-9 - 14y = 5$$
$$-14y = 14$$
$$y = -1$$

Don't wander off before you finish the job. You have to find the value of x, too. Go back to either one of the original equations.

$$x + 4y = -3$$
$$x + 4(-1) = -3$$
$$x - 4 = -3$$
$$x = 1$$

The solution of the system is $x = 1$, $y = -1$.

CHECK POINT

Solve each system by substitution. Check your answers in both equations.

11. $x - y = -9$
    $3x + 2y = 23$

12. $y = 2x - 7$
    $4x + y = 8$

13. $5x + y = 12$
    $3x - 7y = 30$

14. $2x - y = 1$
    $7x - 3y = 1$

15. $b = 8 - 2a$
    $3a + 2b = 15$

16. $2x + 3y = 21$
    $4x + y = 9$

17. $4x = 3y + 14$
    $2y = 19 - 3x$

18. $5x - 2y = 29$
    $3x + 4y = 33$

19. $2x + 4y = 6$
    $4x + 5y = 6$

20. $2a + 3b = 5$
    $3a - 8b = -30$

The substitution method has some significant advantages over graphing, especially when large numbers or fractions are involved. However, there are situations in which substitution gets messy, too. There's another method, also based on the techniques you use for solving equations, that will help solve many systems easily.

Solving by Linear Combinations

The method of linear combinations is sometimes called the elimination method, because it focuses on eliminating a variable. Truthfully, the substitution method eliminates a variable, too. The combination method does it by adding or subtracting, rather than substituting.

You know that when solving an equation, you can add the same amount to both sides and the equation will stay in balance. The linear combination method gets a little bit clever with that fact. Suppose you're looking for two numbers that add to 59 and differ by 25. You call the larger number x and the smaller one y, and your problem can be represented by this system of equations.

$$x + y = 59$$
$$x - y = 25$$

You could take the top equation and add a constant to both sides. You could add 25 to both sides, but it wouldn't really do much.

$$x + y = 59$$
$$x + y + 25 = 59 + 25$$
$$x + y + 25 = 84$$

You could add $-x$ to both sides, but that would just rearrange things.

$$x + y = 59$$
$$x + y - x = 59 - x$$
$$y = 59 - x$$

You could even add $x - y$ to both sides, but that really wouldn't help.

$$x + y = 59$$
$$x + y + x - y = 59 + x - y$$
$$2x = 59 + x - y$$

So here comes the clever part. You're allowed to add $x - y$ to both sides, and you're allowed to add 25 to both sides, but according to the second equation, $x - y$ and 25 are the same amount. They're equal. $x - y = 25$. So you're going to add $x - y$ to one side of the top equation and 25 to the other. It's okay because $x - y$ is just another name for 25.

$$x + y = 59$$
$$x - y = 25$$
$$x + y + x - y = 59 + 25$$
$$2x = 84$$
$$x = 42$$

Doing that makes the y's disappear, leaving us with just one variable, so we can solve for x, then go back and find y.

$$x = 42$$
$$42 + y = 59$$
$$y = 59 - 42$$
$$y = 17$$

This method of solving a system is sometimes just called the addition method, but it takes its official name from the fact that we're combining the two equations in a way that eliminates one variable.

 TIP

If you combine the two equations and both variables disappear, and what's left is true, like $0 = 0$ or $42 = 42$, the system has infinitely many solutions. If both variables disappear and what's left is false, like $0 = 4$ or $8 = -3$, then the system has no solution.

You may wonder if it was just a coincidence that adding the two equations eliminated a variable. That doesn't happen all the time, does it? The simple answer is no, it doesn't, but you can make some adjustments in order to make it happen.

When you look at a system and you want to solve by linear combination, take a step at a time.

1. Simplify the two equations and arrange them so the x-terms, the y-terms, the constants, and the equal signs line up. This is not what you want.

    $$3x + 4y = 25$$
    $$x = 2y - 5$$

    But you can rewrite it like this.

    $$3x + 4y = 25$$
    $$x - 2y = -5$$

2. Ask yourself if adding the equations together will eliminate a variable. If you add the like terms that are now lined up, and you still have two variables, you need to try something else. This is not good.

    $$3x + 4y = 25$$
    $$\underline{x - 2y = -5}$$
    $$4x + 2y = 20$$

3. Check to see if subtracting the equations will eliminate a variable. Subtracting won't do the job here either, but don't give up.

    $$3x + 4y = 25$$
    $$\underline{x - 2y = -5}$$
    $$2x + 6y = 30$$

4. Focus on one variable at a time, and ask yourself whether there is a common multiple for the two coefficients. In the example, $3x$ and x have coefficients of 3 and 1, which have a common multiple of 3. $4y$ and $-2y$ could use 4 or -4. Don't worry about signs just now. Let's just say 4.

If you can find a convenient common multiple for one or both variables, you can still make the linear combination method work. You do it by multiplying the equations—one or both—by constants. You know you can multiply both sides of an equation by any non-zero constant without unbalancing the equation. With your equations, you can choose which variable to pay attention to, and that will determine what constant you multiply by.

If you focus on the x-terms, you can multiply the bottom equation by 3.

$$3x + 4y = 25$$
$$3(x - 2y) = (-5) \cdot 3$$

This will give you a system in which the x-terms have the same coefficient. Subtracting the equations will eliminate the x-terms but be careful to subtract all the way through.

$$
\begin{array}{l}
3x + 4y = 25 \\
\underline{3x - 6y = -15} \\
10y = 40 \\
y = 4
\end{array}
\qquad \Rightarrow \qquad
\begin{array}{l}
3x - 6y = -15 \\
3x - 6 \cdot 4 = -15 \\
3x - 24 = -15 \\
3x = 9 \\
x = 3
\end{array}
$$

If you choose instead to focus on the y-terms, you should multiply the bottom equation by 2. That will give you a system in which you can eliminate the y-terms by adding the equations.

$$
\begin{array}{l}
3x + 4y = 25 \\
2(x - 2y) = (-5) \cdot 2
\end{array}
\qquad \Rightarrow \qquad
\begin{array}{l}
3x + 4y = 25 \\
2x - 4y = -10
\end{array}
$$

Once you add, you can solve for x, and then for y.

$$
\begin{array}{l}
3x + 4y = 25 \\
\underline{2x - 4y = -10} \\
5x = 15 \\
x = 3
\end{array}
\qquad \Rightarrow \qquad
\begin{array}{l}
3x + 4y = 25 \\
3 \cdot 3 + 4y = 25 \\
9 + 4y = 25 \\
4y = 16 \\
y = 4
\end{array}
$$

The choice of which variable to eliminate is yours. You only have to eliminate one variable, and you can choose whichever one you think is easier.

 TIP

When choosing which variable to eliminate, think about the signs. If the terms with a certain variable have opposite signs, you'll be able to eliminate by adding. If they have the same sign, you'll have to subtract, and errors are more common when subtracting. If you'd like to change the sign of a term, you can, but only by changing the signs of all terms in that equation. Multiply through by a negative number and all the signs will reverse.

Let's work on solving this system by linear combinations.

$$3x + 4y = -1$$
$$2x + 5y = -3$$

If you want the x-terms to match, you'll need to multiply the top equation by 2 and the bottom one by 3.

$$
\begin{array}{l} 3x + 4y = -1 \\ 2x + 5y = -3 \end{array} \Rightarrow
\begin{array}{l} 2(3x + 4y) = (-1) \cdot 2 \\ 3(2x + 5y) = (-3) \cdot 3 \end{array} \Rightarrow
\begin{array}{l} 6x + 8y = -2 \\ 6x + 15y = -9 \end{array}
$$

If you'd rather eliminate the y-terms, you'd need to multiply the top equation by 5 and the bottom one by 4.

$$
\begin{array}{l} 3x + 4y = -1 \\ 2x + 5y = -3 \end{array} \Rightarrow
\begin{array}{l} 5(3x + 4y) = (-1) \cdot 5 \\ 4(2x + 5y) = (-3) \cdot 4 \end{array} \Rightarrow
\begin{array}{l} 15x + 20y = -5 \\ 8x + 20y = -12 \end{array}
$$

Once you have a system in which one pair of variable terms have either exactly the same coefficients or exactly opposite coefficients, you can eliminate a variable by either adding or subtracting. If the coefficients are identical, subtract. If they have opposite signs, add. In this example, you'll need to subtract.

$$
\begin{array}{l} 3x + 4y = -1 \\ 2x + 5y = -3 \end{array} \Rightarrow
\begin{array}{l} 5(3x + 4y) = (-1) \cdot 5 \\ 4(2x + 5y) = (-3) \cdot 4 \end{array} \Rightarrow
\begin{array}{l} 15x + 20y = -5 \\ \underline{8x + 20y = -12} \end{array}
$$

$$7x = 7$$
$$x = 1$$
$$3x + 4y = -1$$
$$3 \cdot 1 + 4y = -1$$
$$3 + 4y = -1$$
$$4y = -4$$
$$y = -1$$

CHECK POINT

Use linear combinations to solve each system. Check your answer in both equations.

21. $x + y = 17$
    $x - y = 3$

22. $x - 2y = 5$
    $3x + 2y = 23$

23. $2x + y = 7$
    $8x + y = 19$

24. $7x + 5y = 21$
    $7x + 9y = 21$

25. $5x + y = 12$
    $3x - 7y = 30$

26. $4x + 3y = 7$
    $2x - 5y = 23$

27. $5x - 2y = 29$
    $3x + 4y = 33$

28. $4x - 3y = 14$
    $3x + 2y = 19$

29. $3x - 2y = 1$
    $2x + 3y = 18$

30. $4x + 5y = 6$
    $2x + 4y = 6$

Applications of Systems

Systems of linear equations can be solved by graphing, by substitution, or by linear combination. The need to solve a system usually arises from a problem in which there is more than one unknown. Many of these problems are mixture problems. It may be that substances are being mixed, for a package of mixed nuts or candies, or the chemicals in a lab. It may be a collection of coins of different values, or investments at different interest rates.

To solve any sort of mixture problem that needs two variables, you can follow the same basic pattern. Start by defining what unknown quantity each variable stands for. Then ask yourself, and answer, a few questions. How much, or how many, of each substance do we have? How many pounds of nuts, or how many liters of a chemical, or how many of that type of coin? Then, what is the price per pound, or the concentration of the active ingredient, or the value of the coin?

The total value of the substance can usually be found by multiplying how much you have by the unit value. If you have 3 pounds of nuts that sell for $8 a pound, they are worth $3 \times \$8 = \$24$. Seven dimes, coins worth 10 cents each, are worth 70 cents. If a chemical is 80% acid and you have x liters of it, $80\% \cdot x = 0.8x$ liters are actually acid. It may be helpful to organize the information in a chart like this one.

Type of substance?	How much or how many?	What is one worth?	What is the total value?

Let's look at an example. A collection of nickels and dimes, 16 coins in all, is worth $1.25. How many nickels and how many dimes does the collection contain?

Start by defining variables. Let's use n for the number of nickels and d for the number of dimes. Your chart would look something like this. The last row will tell you about the total collection.

Type of substance?	How much or how many?	What is one worth?	What is the total value?
nickels	n	$0.05	0.05n$
dimes	d	$0.10	0.10d$
collection	16	Can't say	$1.25

You can get your two equations from the table. One equation is produced by adding down the how much column. $n + d = 16$. The other equation comes from adding down the total value column. $0.05n + 0.10d = 1.25$ To solve the problem, you need to solve the system below.

$$n + d = 16$$
$$0.05n + 0.10d = 1.25$$

If you multiply the bottom equation by 10, you'll be able to eliminate the d by subtracting.

$$\begin{aligned} n + d &= 16 \\ 10(0.05n + 0.10d) &= (1.25) \cdot 10 \end{aligned} \quad \Rightarrow \quad \begin{aligned} n + d &= 16 \\ 0.5n + d &= 12.5 \end{aligned}$$

Then you can solve the system.

$$\begin{aligned} n + d &= 16 \\ \underline{0.5n + d = 12.5} \\ 0.5n &= 3.5 \\ n &= 7 \end{aligned} \quad \Rightarrow \quad \begin{aligned} n + d &= 16 \\ 7 + d &= 16 \\ d &= 9 \end{aligned}$$

Your solution says there are 7 nickels, worth a total of 35 cents, and 9 dimes, worth a total of 90 cents. Sure enough, 35 cents plus 90 cents is $1.25.

Another common type of situation that leads to a system is the distance problem, which talks about two vehicles moving at different speeds. The basic rule for this sort of problem is distance equals rate times time. The total distance covered equals the rate of speed multiplied by the time spent traveling at that speed.

Suppose you need to get to a town 138 miles from your home. You start out on foot, walking at 4 miles per hour, and you walk until you get to the bus stop. You get on the bus, and travel at 28 miles per hour, until you reach your destination, exactly 6 hours after you left home. How long did you walk?

Let's define variables. Let x represent the number of hours you walked. What's your second variable? You know your walking speed and the speed of the bus, so it must be the time on the bus. Let y be the number of hours you spent on the bus. You can use a chart here, too.

Mode of travel	Rate of Speed	Time at that speed	Total distance covered
Walking	4	x	$4x$
Bus	28	y	$28y$
Total trip	Can't say	6	138

You can generate one equation by adding down the time column. $x + y = 6$. Then you can get a second equation by adding down the distance column. $4x + 28y = 138$. To solve the problem, solve the system.

$$x + y = 6$$
$$4x + 28y = 138$$

Multiply the top equation by 4 to eliminate the x's or 28 to eliminate the y's. Let's multiply by 4, and then subtract.

$$\begin{aligned} 4x + 4y &= 24 \\ 4x + 28y &= 138 \\ \hline -24y &= -114 \\ y &= 4\frac{3}{4} \end{aligned}$$

$$\Rightarrow \quad \begin{aligned} x + y &= 6 \\ x + 4\frac{3}{4} &= 6 \\ x &= 1\frac{1}{4} \end{aligned}$$

The solution is $x = 1\frac{1}{4}$ and $y = 4\frac{3}{4}$, which means you spent 1 hour and 15 minutes walking and 4 hours and 45 minutes on the bus.

CHECK POINT

Solve each problem using a system of equations. Use whatever method seems most efficient.

31. A collection of 20 coins, all dimes and quarters, totals $3.05. How many dimes and how many quarters are in the collection?

32. A stamp collector bought 30 stamps. Some stamps cost 14 cents and some cost 17 cents. The total price was $4.56. How many of each stamp did he buy?

33. A coffee merchant wants to mix coffee worth $6.00 a pound with coffee worth $5.10 a pound, to create a blend that she can sell for $5.40 a pound. She wants to produce a total of 60 pounds of the new blend. How many pounds of each type of coffee should she use?

34. I had $6,000 to invest. I invested part of it in a fund that pays 4% interest and the rest in a fund that pays 7% interest. If the total interest from the two accounts was $342, how much did I invest at each rate?

35. You need 48 ml of a 15% salt solution. You have a 5% salt solution and a 20% salt solution. How much of each should you combine to make 48 ml of a 15% solution?

36. A driver began a trip of 352 miles traveling at 50 miles per hour. After a period of time, she had to slow to 20 miles per hour because of construction, and completed the trip at that speed. If her total travel time was 8 hours, how long did she travel at each speed?

Define your variables, then translate the problem into two equations. Solve and check.

37. A group of 8 adults and 10 children paid a total of $115 for admission to an amusement park. A group of 3 adults and 4 children paid a total of $44 for admission to the same park. Find the price of admission for an adult, and the admission price for a child.

38. A number of nickels and dimes have a total value of $1.65. If the number of nickels and dimes were interchanged, the total value would be $1.35. How many nickels and how many dimes are there in the original collection?

39. An order of 3 pens and 5 pencils costs $3.90. In the same shop, an order of 5 pens and 2 pencils costs $4.60. find the price of 1 pen and the price of 1 pencil.

40. The attendance at a carnival was 250 people, some adults, some children. The total collected was $735. The adult admission price was $3.50 and the child's admission was $2.50. How many adults and how many children attended the carnival?

We often find ourselves in situations where we're searching for more than one piece of information. Being able to solve a system of equations is an important tool for those problems. If you can represent the quantities you're looking for by variables and you have enough information to write two equations involving those variables, you can solve to find them. But life doesn't always cooperate by giving you two nice tidy equations. Sometimes our information comes in the form of inequalities.

Systems of Inequalities

Imagine that you're in charge of organizing a party. You have a budget for food and you know how much it will cost to feed an adult and how much it will cost to find a child. The number of people you can invite is limited by the size of your home, but there is also a minimum number, because certain people must be invited.

This situation sounds as though it might be represented by a system of equations. Let x represent the number of adults you invite and y represent the number of children. When you start to write about the situation, however, you may realize that you don't have to spend every penny of your budget. It would be okay to spend less than what you budgeted, as long as all your needs were covered. And there may not be a solution of the form "invite this many adults and that many children." Instead, you'll probably end up with a range of possibilities. You're dealing with an inequality situation.

TIP

The choice between a system of equations and a system of inequalities is not always clear, but if you're interested in all the possibilities, you probably want a system of inequalities. If you just want to know about the dividing line between more and less, or acceptable and unacceptable, you can probably use a system of equations.

Writing a set of inequalities to describe a situation is very much the same as writing equations, but only the graphing method lets you really see your solution, so that's how systems of inequalities are tackled. The other difference between systems of equations and systems of inequalities is that you may find that you have more than two inequalities. Each inequality sets a boundary on the acceptable values in the solution set.

Let's put some numbers into the party situation and see what it looks like. Let's say you know it costs $5 for food for an adult and $3 for a child and your budget is $100. You must invite the 6 people in your immediate family at a minimum, and you don't think your home can hold more than 25 people, so that's your maximum.

Using x for the number of adults and y for the number of children, you would have three inequalities. (The last two could be combined into a compound inequality.)

$$5x + 3y \leq 100$$
$$x + y \geq 6$$
$$x + y \leq 25$$

TIP

When graphing a system of inequalities, especially one with more than two equations, don't rush to shade entire half-planes. You can wind up with so much shading that you can't see where the overlap is. Instead, try putting a narrow fringe of shading on the correct side of the line. After you've done that for all your inequalities, look for the region that has fringe on all sides.

If you graph all those inequalities on one axis, you'll get the following graph. You only need to look at Quadrant I because you won't be inviting a negative number of people.

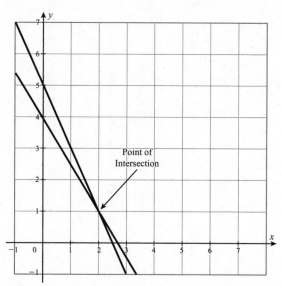

All that shading can be hard to understand, so it's worthwhile to focus on the portion that is shaded by all three inequalities and mark it clearly.

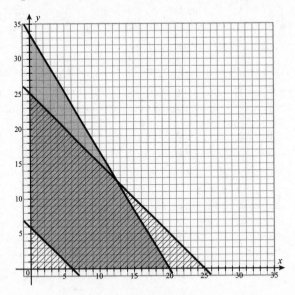

Any point in that shaded region represents a number of adults and a number of children that you could invite while staying within the limitations you have.

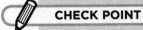 **CHECK POINT**

Graph each system of inequalities and clearly mark the area that represents the solution.

41. $y \leq 2x + 1$
    $y \geq 4 - x$

42. $y < 7 - 3x$
    $2x + y > 8$

43. $3x - y < 1$
    $y \geq 2 - 3x$

44. $5x + 2y \geq 12$
    $3x + 7 < y$

45. $3x + 4 \geq 4y$
    $x + 2y < 12$

Use the graph of a system of inequalities to show the solution set for each problem.

46. The sum of two numbers is at least 40 and their difference is no more than 8. What are the numbers?

47. Find two numbers whose difference is no more than 12 and whose sum is at least 20.

48. A small airline uses planes that can seat no more than 80 passengers, some first class, some coach. For a certain flight, they charge $100 for a coach ticket and $300 for a first class seat. If the airline wants to collect at least $7500 in fares for the flight, how many first class and how many coach seats should they provide?

49. Gillian creates metal sculptures from nuts and bolts, and has decided to produce some of her more popular creations in quantity for sale to gift shops. Her smallest sculpture, Gull, uses 15 nuts and 12 bolts. Her largest piece, Giant, is built from 20 nuts and 30 bolts. Gillian has 400 nuts and 300 bolts available for sculpting. How many of each sculpture can Gillian produce?

50. A small bakery sells cookies and cakes, and bakes large batches of each, using flour and sugar in each recipe. Each batch of cookies uses 2 pounds of flour and 3 cups of sugar. Each batch of cakes uses 9 pounds of flour and 4 cups of sugar. On any day, there are 25 pounds of flour and 36 cups of sugar available. How many batches of cookies and how many batches of cakes can they produce in a day?

The Least You Need to Know

- A pair of linear equations using the same two variables that are solved simultaneously are called a linear system.

- You can solve a system by graphing both equations and finding the point of intersection.

- You can solve a system by isolating a variable in one equation and using the result to substitute for that variable in the other equation.

- You can solve a system by combining the equations, or multiples of the equations, by addition or subtraction to eliminate one variable.

- Once the value of one variable has been found, you can substitute that variable into one of the equations to solve for the other variable.

Polynomials

My detective stories start out with everything seeming clear, but before the matter gets resolved, you know it's going to get complicated. What do the heroes do when things get complex? Usually, they step back and try to apply essential truths to the evidence they've gathered, and often they actually have to back up. They need to work backward to understand what they're dealing with. Once they've navigated those complications, however, they always find the critical piece of evidence that unravels the entire mystery.

With Part 4, we've reached that point in our algebra story. Polynomials look complicated at first glance, but we'll master them by applying crucial rules we learned in arithmetic. When we've learned to operate on those polynomials, we'll do our backing up. We'll factor polynomials, taking them back to the products that produced them. That will be the key that makes everything that follows fall into place. Isn't that what always happens?

Adding and Subtracting Polynomials

In previous chapters, we've talked a lot about terms, expressions, and equations. That's natural, because the language of algebra is built on those elements. In this chapter, and several that follow, we're going to focus on certain types of terms, expressions, and equations. We're going to bring our focus to a category of algebraic operations, working with expressions called polynomials. Much of what you've learned will carry over, but polynomials will present some new concerns. As is so often the case, we'll begin with some new vocabulary.

Form and Degree

A *monomial* is a product of a real number, called the coefficient, and a variable raised to a non-negative integer power. Take that apart and you'll see that monomial is the name given to the product of a coefficient and a power of a variable. Examples of monomials would include $-\dfrac{2}{3}x^2$ and $4t^5$.

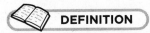

DEFINITION

A **monomial** is a constant, a variable, or a product of constants and variables.

If you're thinking that sounds a lot like what we've been calling a term, you're right. A monomial is a term, with a few additional properties. The exponents cannot be negative numbers and cannot be fractions. An exponent of 1 is fine, so $-3x$ is a monomial, and an exponent of 0 is acceptable, too, so a constant like $-7 = -7x^0$ also counts as a monomial. But if there's a variable in a denominator, like $\frac{4}{x}$, or under a radical, like $\sqrt{2x}$, it's not a monomial.

Most of the monomials you'll encounter involve only one variable, although it is possible to have a monomial with more than one variable. The *degree* of a monomial is the power to which the variable is raised. $-\frac{2}{3}x^2$ is degree two, or second degree. $4t^5$ is fifth degree. $-3x$ is first degree, degree 1, and constants are degree zero.

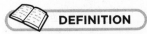

DEFINITION

The **degree** of a monomial with one variable is the power of the variable. A constant has degree zero.

When you add or subtract monomials, you form expressions that take the general name *polynomials*. This is the general name because it applies to any expression that is made up of monomials, including just one monomial. A monomial is a polynomial, and any sum of monomials is a polynomial. $6x^3$ and $-9x^8$ are both monomials, and $6x^3 + -9x^8$ or $6x^3 - 9x^8$ is a polynomial. The expression $x^6 + 4x^5 - 7x^4 + 5x^3 + \frac{1}{2}x^2 - 2x + 1$ is also a polynomial. The term polynomial applies no matter how many monomials are involved.

The polynomials you will work with most often generally have two or three monomials, and those common ones get their own names. A polynomial made of two monomials is a *binomial*, and one with three monomials is a *trinomial*. Anything bigger than that gets the general name polynomial.

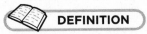

DEFINITION

A **polynomial** is a sum of monomials.

To determine the degree of a polynomial, look at the degree of each of the monomials that form the polynomial. The highest of those is the degree of the polynomial. $x^6 + 4x^5 - 7x^4 + 5x^3 + \frac{1}{2}x^2 - 2x + 1$ is a sixth-degree polynomial and $6x^3 + -9x^8$ is an eighth-degree binomial. You will most often see first-degree binomials like $7x - 3$ and second-degree trinomials like $x^2 + 4x - 9$.

It's easier to determine the degree of a polynomial if it's in standard form. Standard form means that the monomials that make up the polynomial have been arranged in order, from highest degree to lowest. $x^6 + 4x^5 - 7x^4 + 5x^3 + \frac{1}{2}x^2 - 2x + 1$ and $x^2 + 4x - 9$ are both polynomials in standard form. When a polynomial is in standard form, the degree of the polynomial is the degree of the first term.

CHECK POINT

Tell whether the expression is a polynomial.

1. $x^2 + 4x - 7 + \dfrac{3}{x}$

2. $3y^7 - 4y^3 + 8$

3. $2t^3 + 4t^2 - 8t + 4\sqrt{t}$

Put each polynomial in standard form.

4. $9z - 6z^2 + 2$

5. $4x^2 - 3x + x^3 - 2$

6. $8t - 2 + t^2$

Identify each expression as a monomial, binomial, trinomial, or polynomial, and give its degree.

7. $9y - 3$

8. $x^2 + 5x - 9$

9. $-\dfrac{1}{2}z^3$

10. $12a^9 + 8a^{12} - 4a^7 + 6a^4 + 2$

Addition of Polynomials

You already know most of the rules you need to do arithmetic with polynomials. Because polynomials are a class of expressions, the rules you learned for operations with variables will apply.

To add two polynomials, you simply need to combine like terms. If you want to add the binomial $-3x - 6$ to the trinomial $4x^2 - 7x + 12$, your job will be to add the like terms and only the like terms. Both of these polynomials are in standard form, so it will be easy to write them one under another, with like terms aligned. This vertical arrangement makes addition easier.

$$4x^2 - 7x + 12$$
$$\underline{\quad\quad -3x - 6}$$

From here, it's basically the same process used in standard arithmetic. When you add two numbers, you first add the ones digits, then the tens digits, and so on, right to left. When you add two polynomials, you first add the constants, then the first degree terms, and so on, right to left. There's no need to carry. Just add straight down.

$$
\begin{array}{r}
4x^2 - 7x + 12 \\
-3x - 6 \\
\hline
4x^2 - 10x + 6
\end{array}
$$

The vertical arrangement is convenient when the polynomials are in standard form and particularly useful when adding more than two polynomials. When you add polynomials vertically, you just leave an empty space if a particular power is absent from a polynomial. If you add $3x^6 + 9x^4 - 7x^3 + 2x^2 - 1$ and $4x^5 - 3x^4 + 11x^2 - x + 8$, you place them one under the other with terms that contain the same power of the variable aligned. Space the terms out whenever a power is missing, so that the terms in each column are like terms.

$$
\begin{array}{ccccccc}
3x^6 & & +9x^4 & -7x^3 & +2x^2 & & -1 \\
& 4x^5 & -3x^4 & & +11x^2 & -x & +8 \\
\hline
\end{array}
$$

If you prefer, you can insert zeros to hold places for the missing powers.

$$
\begin{array}{ccccccc}
3x^6 & +0x^5 & +9x^4 & -7x^3 & +2x^2 & +0x & -1 \\
& 4x^5 & -3x^4 & +0x^3 & +11x^2 & -x & +8 \\
\hline
\end{array}
$$

Once the polynomials are arranged with like terms in columns, you add down the column. If there's only one term in a column, you can carry it down.

$$
\begin{array}{ccccccc}
3x^6 & & +9x^4 & -7x^3 & +2x^2 & & -1 \\
& 4x^5 & -3x^4 & & +11x^2 & -x & +8 \\
\hline
3x^6 & +4x^5 & +6x^4 & -7x^3 & +13x^2 & -x & +7
\end{array}
$$

If there are more than two polynomials to be added, you still add down each column, you just need a taller stack. To add $6x^2 - 3x + 2$, $9x^2 - 7x + 1$, and $4x^2 + 9x - 3$, stack them with like terms aligned.

$$
\begin{array}{r}
6x^2 - 3x + 2 \\
9x^2 - 7x + 1 \\
4x^2 + 9x - 3 \\
\hline
\end{array}
$$

The columns are taller, but the process is the same. Add two terms, and then add the third term to that total.

$$6x^2 - 3x + 2$$
$$9x^2 - 7x + 1$$
$$\underline{4x^2 + 9x - 3}$$
$$19x^2 - x + 0$$

You can drop the +0 and just give the sum as $19x^2 - x$.

If the polynomials are not in standard form, or they're not too complicated, you might prefer not to use a vertical arrangement, and just keep all your work on a line. If you do make that choice, organization will be important. Either take a moment to rearrange the terms so that like terms are grouped together, or cross off terms as you add them.

If you're adding $4t^3 - 7t + 9t^2 + 2$ to $6 - 2t + t^3 - 8t^4 + t^2$, you might want to organize the problem this way.

$$\left(4t^3 - 7t + 9t^2 + 2\right) + \left(6 - 2t + t^3 - 8t^4 + t^2\right)$$
$$= -8t^4 + \left(4t^3 + t^3\right) + \left(9t^2 + t^2\right) + \left(-7t - 2t\right) + \left(2 + 6\right)$$
$$= -8t^4 + 5t^3 + 10t^2 - 9t + 8$$

 ALGEBRA TRAP

The biggest danger in rearranging the terms of the polynomials is separating a number from its sign. Keep every sign firmly attached to a coefficient. If you move a term, make sure its sign moves with it.

If you don't want to rewrite the problem to rearrange the terms, you can cross out terms to keep track of what you've done.

$$\left(4t^3 - 7t + 9t^2 + 2\right) + \left(6 - 2t + t^3 - 8t^4 + t^2\right)$$

There's only one fourth power term, so write that down and cross it out.

$$\left(4t^3 - 7t + 9t^2 + 2\right) + \left(6 - 2t + t^3 \cancel{-8t^4} + t^2\right) = -8t^4$$

Add the third power terms and cross them out when you're done.

$$\left(\cancel{4t^3} - 7t + 9t^2 + 2\right) + \left(6 - 2t \cancel{+t^3} \cancel{-8t^4} + t^2\right) = -8t^4 + 5t^3$$

Next the squares, then the first power terms.

$$\left(\cancel{4t^3}\ \cancel{-7t}\ \cancel{+9t^2}+2\right)+\left(6\ \cancel{-2t}\ \cancel{+t^3}\ \cancel{-8t^4}\ \cancel{+t^2}\right)=-8t^4+5t^3+10t^2-9t$$

Combine the constants and you're done.

$$\left(\cancel{4t^3}\ \cancel{-7t}\ \cancel{+9t^2}\ \cancel{+2}\right)+\left(\cancel{6}\ \cancel{-2t}\ \cancel{+t^3}\ \cancel{-8t^4}\ \cancel{+t^2}\right)=-8t^4+5t^3+10t^2-9t+8$$

When every term is crossed out, you know your job is done.

CHECK POINT

Perform each addition. Use the method you think is most efficient.

11. $\left(5x^3\right)+\left(7x^2\right)+\left(2x^3\right)$

12. $\left(4x-7\right)+\left(9-2x\right)$

13. $\left(3b+b^2\right)+\left(5b-6b^2\right)$

14. $\left(3x^2-2x-5\right)+\left(-5x^2-4x-6\right)$

15. $\left(x^2-3x+4\right)+\left(-7x^2-5x\right)$

16. $\left(x^2-8\right)+\left(5x^2+6x-1\right)$

17. $\left(2y^3+3y^2-y\right)+\left(5y^3-6y^2+3y\right)+\left(2y^3-y^2-7y\right)$

18. $\left(2y^2+6y-5\right)+\left(7y-3y^2+6\right)+\left(8-9y+2y^2\right)$

19. $\left(3t^2-2t+6\right)+\left(7t-t^2-1\right)+\left(4-6t^2+t\right)$

20. $\left(2x+3x^2\right)+\left(-5x^3-4x\right)+\left(7x^3-x^2\right)$

Subtraction of Polynomials

You know that subtraction can be treated as adding the opposite. To find the opposite of a number, change the sign. Finding the opposite of a term just requires changing the sign of the coefficient. A polynomial is a sum of monomials, and each monomial is a term. To find the opposite of a monomial change the sign of its coefficient., Finding the opposite of a polynomial will mean changing all the signs. The opposite of $4t^3-7t+9t^2+2$ will be $-4t^3+7t-9t^2-2$. Remember that the opposite is what you add to get to zero.

$$\left(4t^3-7t+9t^2+2\right)+\left(-4t^3+7t-9t^2-2\right)=\left(4t^3-4t^3\right)+\left(-7t+7t\right)+\left(+9t^2-9t^2\right)+\left(2-2\right)=0$$

Forming the opposite of a polynomial by changing the sign of each of its monomials is equivalent to multiplying the polynomial by -1. The opposite of $3x^6 + 9x^4 - 7x^3 + 2x^2 - 1$ is $-1(3x^6 + 9x^4 - 7x^3 + 2x^2 - 1)$, and according to the distributive property, that equals $(-1)(3x^6) + (-1)(9x^4) + (-1)(-7x^3) + (-1)(2x^2) + (-1)(-1) = -3x^6 - 9x^4 + 7x^3 - 2x^2 + 1$. This is why we sometimes use the expression "distribute the negative." It's actually a -1, but the effect is to change all the signs.

To subtract polynomials, you add the opposite of the second polynomial to the first polynomial. To perform the subtraction below, first rewrite it as adding the opposite.

$$(4x^3 - 8x^2 + 3x - 9) - (6x^3 + 2x^2 - 7x + 6)$$
$$(4x^3 - 8x^2 + 3x - 9) + (-6x^3 - 2x^2 + 7x - 6)$$

Once you've changed to addition of the opposite, you can just go ahead and add as you've done before, either on a line or in vertical format. Here's this example on a line.

$$(4x^3 - 8x^2 + 3x - 9) - (6x^3 + 2x^2 - 7x + 6)$$
$$(4x^3 - 8x^2 + 3x - 9) + (-6x^3 - 2x^2 + 7x - 6)$$
$$(4x^3 - 6x^3) + (-8x^2 - 2x^2) + (3x + 7x) + (-9 - 6)$$
$$-2x^3 - 10x^2 + 10x - 15$$

Let's do the subtraction $(y^5 + 4y^3 - 9y^2 + 3) - (8y^4 - 2y^3 + 5y - 7)$ in vertical format. It may show another reason that adding the opposite is the best course of action. If you think of it as a subtraction problem, you face an awkward question once you line up like terms: what do you take $5y$ away from? There's no term there.

$$
\begin{array}{cccccc}
y^5 & & +4y^3 & -9y^2 & & +3 \\
\underline{-\left(\quad 8y^4 \quad -2y^3 \quad \quad +5y \quad -7\right)} \\
\end{array}
$$

The actual answer is that you're subtracting $0 - 5y$, but switching to adding the opposite removes the question. If you change to the opposite of the bottom polynomial, you can just add down each column.

$$
\begin{array}{cccccc}
y^5 & & +4y^3 & -9y^2 & & +3 \\
& -8y^4 & +2y^3 & & -5y & +7 \\
\hline
y^5 & -8y^4 & +6y^3 & -9y^2 & -5y & +10 \\
\end{array}
$$

ALGEBRA TRAP

When subtracting, it's easy to make sign errors. Take the time to change all the signs of the bottom polynomial and then add. If you try to change signs as you go, it's easy to change one or two and then forget to change the later ones.

CHECK POINT

Perform each subtraction by adding the opposite of the second polynomial.

21. $(8x-3)-(4-7x)$

22. $(9x^2-7x+5)-(-2x^2+6x+3)$

23. $(2x^2-3x)-(4x^2+5x-2)$

24. $(2y^2+7y-4)-(3y^2-5y+2)$

25. $(5t^2-7t+2)-(3t^2-9t-1)$

26. $(2a^2+6-3a)-(8+9a-a^2)$

27. $(5x^2-2x+3)-(2x^2-x+5)$

28. $(y^3+5)-(y^2-3)$

29. $(4x^3+5x^2-7)-(2x^2-8x-1)$

30. $(2x^2+3x)-(5x^2+2)$

Combined Operations

For longer problems that combine addition and subtraction, you don't need any new procedures, but it is especially important for problems like these that you change subtraction to addition of the opposite. Trying to shift operations as you're working will just set you up to make mistakes. Get everything over to addition, and then add all the way through. Let's look at a couple of examples.

$$(3x-7)-(x^2+4)+(5x^2+2x-3)-(x^2-2x-1)$$

Change each subtraction to addition of the opposite.

$$(3x-7)+(-x^2-4)+(5x^2+2x-3)+(-x^2+2x+1)$$

Rearrange to group like terms.

$$(-x^2+5x^2-x^2)+(3x+2x+2x)+(-7-4-3+1)=3x^2+7x-13$$

One of the advantages of changing to adding the opposite is that it allows you to use a vertical alignment. You can add several polynomials in vertical arrangement but you can't do that if you have some addition and some subtraction. Here's an example.

$$\left(x^2+7x-3\right)-\left(2x^2-3x-1\right)+\left(5x^2+8x\right)-\left(2x-7\right)$$

Change to adding the opposite.

$$\left(x^2+7x-3\right)+\left(-2x^2+3x+1\right)+\left(5x^2+8x\right)+\left(-2x+7\right)$$

Now you can stack in vertical format.

$$
\begin{array}{rrr}
x^2 & +7x & -3 \\
-2x^2 & +3x & +1 \\
5x^2 & +8x & \\
& -2x & +7 \\
\hline
4x^2 & +16x & +5
\end{array}
$$

CHECK POINT

Combine the polynomials by adding or subtracting as indicated. Give your answer as a polynomial in standard form.

31. $\left(8x-3\right)+\left(4-7x\right)-\left(4x-7\right)+\left(9-2x\right)$

32. $\left(y^3+5\right)+\left(y^2-3\right)+\left(3y+y^2\right)-\left(5y-6y^2\right)$

33. $\left(2x^2-3x\right)-\left(3x^2-2x-5\right)-\left(4x^2+5x-2\right)+\left(-5x^2-4x-6\right)$

34. $\left(x^2-3x+4\right)-\left(-7x^2-5x\right)-\left(2x^2-3x\right)+\left(4x^2+5x-2\right)$

35. $\left(3x^2+2x-1\right)+\left(5x^2-7x+6\right)-\left(7x^2-15x-3\right)$

36. $\left(2x^2+3x\right)-\left(5x^2+2\right)-\left(x^2-8\right)+\left(5x^2+6x-1\right)$

37. $\left(8a^3-4a^2+6a+2\right)-\left(2a^2+6-3a\right)+\left(8+9a-a^2\right)$

38. $\left(4x^3+5x^2-7\right)-\left(7x^3-x^2\right)-\left(2x^2-8x-1\right)+\left(2x+3x^2\right)+\left(-5x^3-4x\right)$

39. $\left(3t^2-9t-1\right)-\left(3t^2-2t+6\right)+\left(4-6t^2+t\right)$

40. $\left(2x^2+3x\right)-\left(-5x^2-4x-6\right)+\left(x^2-2x-5\right)+\left(4x^2-5x-2\right)$

The Least You Need to Know

- Add polynomials by combining like terms.
- Use vertical alignment to organize like terms.
- Form the opposite of a polynomial by changing the sign of every term.
- To subtract a polynomial, add the opposite.
- For combined operations, change all subtractions to adding the opposite, and then add.

Multiplying and Dividing Polynomials

In the last chapter, we looked at how polynomials could fit into our understanding of addition and subtraction. Now it's time to revisit multiplication and division. We need to see what ideas we've already learned that are still useful, and what modifications we might need to make. In this chapter, we'll review methods of multiplication we learned earlier that will still be useful, examine some special cases, and explore what division is possible when working with polynomials.

In This Chapter

- Multiplying a monomial by a polynomial using the distributive property
- Multiplying two binomials using the FOIL rule
- Multiplying larger polynomials in vertical format
- Dividing by a monomial
- Polynomial long division

Multiplying Polynomials

To multiply two polynomials, you'll employ some methods you've already learned. Because each monomial involves a power of the variable, the rule for multiplying powers of the same base will be key. You'll also use the distributive property and the FOIL rule.

To multiply two monomials, like $\left(5x^3\right) \cdot \left(-8x^2\right)$, you multiply coefficient by coefficient and variable by variable. Because each of the variables is raised to a power, you need to use the rule for exponents, and add the exponents.

$$\left(5x^3\right) \cdot \left(-8x^2\right) = 5 \cdot (-8) \cdot \left(x^3 \cdot x^2\right) = -40x^{3+2} = -40x^5$$

If one of the monomial factors is a constant, multiply the coefficients, and the variable factor is unchanged.

$$(-7)\cdot\left(4t^7\right)=(-7)\cdot(4)\cdot\left(t^7\right)=-28t^7$$

If one (or both) of the factors is a first degree monomial, remember that you can write in the exponent of 1 that usually isn't shown, and then apply the exponent rule.

$$\left(-5y^3\right)\cdot\left(7y\right)=\left(-5y^3\right)\cdot\left(7y^1\right)=-35y^{3+1}=-35y^4$$

$$\left(\frac{1}{2}t\right)\cdot(-6t)=\left(\frac{1}{2}t^1\right)\cdot\left(-6t^1\right)=-3t^2$$

Multiplying one monomial by one monomial is the basis of all polynomial multiplication. Each of the methods that we'll explore breaks problems down to a collection of small multiplications. Let's take what we know about multiplying monomials and build to larger problems.

Review of Distributive Property

To multiply a single monomial by a larger polynomial, we're going to turn to our old friend, the distributive property. You've used the distributive property before. It states that multiplying a single term by a sum of terms is equivalent to multiplying that one term by each of the terms in the sum. Applying that to polynomials results in a problem like this one.

$$3x^2\left(5x^3-7x^2+2x-6\right)=\left(3x^2\right)\left(5x^3\right)+\left(3x^2\right)\left(-7x^2\right)+\left(3x^2\right)\left(2x\right)+\left(3x^2\right)(-6)$$
$$=15x^5-21x^4+6x^3-18x^2$$

The distributive property lets you change the problem into a collection of multiplications, all monomial multiplied by monomial. Follow the same rules for those as you have before. Multiply the coefficients and multiply the powers of the variable by keeping the variable and adding the exponents. The distributive property says you can add like terms after multiplying, but if the polynomials were properly simplified before you started, there shouldn't be any like terms to combine.

 THINK ABOUT IT

When you multiply a polynomial by a monomial using the distributive property, each term of the polynomial is multiplied by the monomial. Each of the terms in the polynomial was a different power of the variable, and each was multiplied by the same term. The power of each term was increased by the same amount so the terms of the product will all be distinct powers of the variable, and no combining will be necessary.

You might encounter a larger problem, involving a multiplication as well as other operations, or more than one multiplication, in which you will have simplifying to do at the end. The following example has two different multiplications.

$$4t\left(8t^2 + 3t + 4\right) - 7t^2\left(t^2 - 5t + 2\right)$$

The first multiplication, $4t\left(8t^2 + 3t + 4\right)$, is a monomial multiplied by a trinomial. The trinomial is in simplest form, so after that multiplication, there won't be any like terms.

$$4t\left(8t^2 + 3t + 4\right) = 32t^3 + 12t^2 + 16t$$

The second multiplication, $-7t^2\left(t^2 - 5t + 2\right)$, is also a monomial times a trinomial, with the trinomial in simplest form, so that product has no like terms.

$$-7t^2\left(t^2 - 5t + 2\right) = -7t^4 + 35t^3 - 14t^2$$

When the whole problem gets put together, however, there are terms from the first multiplication that you can combine with terms from the second.

$$4t\left(8t^2 + 3t + 4\right) - 7t^2\left(t^2 - 5t + 2\right)$$
$$32t^3 + 12t^2 + 16t - 7t^4 + 35t^3 - 14t^2$$
$$-7t^4 + \left(32t^3 + 35t^3\right) + \left(12t^2 - 14t^2\right) + 16t$$
$$-7t^4 + 67t^3 - 2t^2 + 16t$$

CHECK POINT

Complete each multiplication using the distributive property.

1. $5a\left(2a^2 + 3a\right)$

2. $-3b^2\left(2b^2 - 3b + 5\right)$

3. $2x^2\left(11x^2 - 3x - 7\right)$

4. $-3x^3\left(2 + 4x - 5x^2\right)$

5. $3y^4\left(2y^6 - y^2\right)$

6. $8x\left(x^2 - 3\right)$

7. $-5a^3\left(2a^4 - 4a^2 + 6\right)$

8. $7x^5\left(9 - 3x^3\right)$

9. $-6x^4\left(x^2 + x + 3\right)$

10. $-x^3\left(x^5 - x^4 + x^3 - x^2 + x - 1\right)$

The other technique that's already part of your toolbox and is a crucial skill for multiplying polynomials is the FOIL rule. It's useful in only one situation, the multiplication of two binomials, but that's a very common event. Multiplying two terms by two terms requires four separate multiplications that are designated as First, Outer, Inner, and Last.

Review of FOIL

The FOIL rule states that $(a+b)(c+d) = ac + ad + bc + bd$ but writing it in that form doesn't really communicate it very well. Let's try this instead. Every binomial is made up of two monomials. Call them the First and the Last, so in the binomial $5x^2 + 3x$, $5x^2$ is the First and $3x$ is the Last. In the binomial $2x^3 - 7x^2$, $2x^3$ is the First and $-7x^2$ is the Last.

When you multiply $(5x^2 + 3x)(2x^3 - 7x^2)$, you have $(\text{First} + \text{Last})(\text{First} + \text{Last})$. You want to multiply First × First and Last × Last, but you also want to multiply Firsts and Lasts together. There's a First × Last pair on the outer ends and a Last × First pair in the middle.

$$\overset{\textit{First}}{\underset{\text{(First + Last)(First + Last)}}{\swarrow \qquad \searrow}}$$

$$\overset{\textit{Outer}}{\underset{\left(\text{First + Last}\right)\left(\text{First + Last}\right)}{\swarrow \qquad\qquad \searrow}}$$

$$\overset{\textit{Inner}}{\underset{\left(\text{First + Last}\right)\left(\text{First + Last}\right)}{\downarrow \qquad \downarrow}}$$

$$\overset{\textit{Last}}{\underset{\left(\text{First + Last}\right)\left(\text{First + Last}\right)}{\swarrow \quad \searrow}}$$

When all the multiplications are done, there may (or may not) be like terms that can be combined. Our earlier example will have like terms.

$$\begin{aligned}
\left(5x^2 + 3x\right)\left(2x^3 - 7x^2\right) &= 5x^2 \cdot 2x^3 + 5x^2\left(-7x^2\right) + 3x \cdot 2x^3 + 3x\left(-7x^2\right) \\
&= 10x^5 - 35x^4 + 6x^4 - 21x^3 \\
&= 10x^5 - 29x^4 - 21x^3
\end{aligned}$$

Here's an example of the multiplication of two binomials that doesn't have any like terms after the multiplication is done.

$$\begin{aligned}
\left(x^2 + 3\right)\left(2x + 4\right) &= x^2 \cdot 2x + x^2 \cdot 4 + 3 \cdot 2x + 2 \cdot 4 \\
&= 2x^3 + 4x^2 + 6x + 8
\end{aligned}$$

The FOIL rule is most often used for multiplying two first-degree binomials to produce a second-degree trinomial. When you multiply $(x+5)(x+2)$, you get a trinomial beginning with x^2.

$$(x+5)(x+2) = x \cdot x + 2 \cdot x + 5 \cdot x + 5 \cdot 2$$
$$= x^2 + 2x + 5x + 10$$
$$= x^2 + 7x + 10$$

The product of $3x-7$ and $-4x+1$ is also a second degree trinomial.

$$(3x-7)(-4x+1) = 3x(-4x) + 3x \cdot 1 + (-7)(-4x) + (-7) \cdot 1$$
$$= -12x^2 + 3x + 28x - 7$$
$$= -12x^2 + 31x - 7$$

CHECK POINT

Complete each multiplication using the FOIL rule.

11. $(y+6)(y-1)$ 16. $(3x-8)(4x+5)$

12. $(x+5)(x+2)$ 17. $(3x+1)(2x-3)$

13. $(t-3)(t-2)$ 18. $(2x-7)(4x+9)$

14. $(b-4)(b+6)$ 19. $(10a-3)(8a-5)$

15. $(y+3)(3y-1)$ 20. $(x^2+9)(x^2-4)$

Special Products

The FOIL rule can handle any product of two binomials, but two common patterns are worth pointing out. One occurs when you multiply a binomial by itself, squaring it, and the other when you multiply the difference of two monomials by their sum.

When you square a binomial, the product is a trinomial. Two of the terms of that trinomial are the squares of each term of the binomial, and the third is twice the product of the two terms. Let's start with a simple example.

$$(x+5)^2 = (x+5)(x+5) = \underbrace{x^2}_{x \text{ squared}} + \underbrace{5x+5x}_{\text{two copies of the product of x and 5}} + \underbrace{25}_{5 \text{ squared}} = x^2 + 10x + 25$$

ALGEBRA TRAP

When squaring a binomial, many people will forget the FOIL rule and just square each term. Don't be one of them. One of the reasons you learn the pattern of the perfect square is so that you won't make that mistake. Each term of the first binomial must be multiplied by both terms of the second. $(a+b)^2 \neq a^2 + b^2$

Recognizing this pattern lets you square binomials quickly. $(t+7)^2 = t^2 + 2(7t) + 7^2 = t^2 + 14t + 49$ or $(y-3)^2 = y^2 + 2(-3y) + (-3)^2 = y^2 - 6y + 9$. The same rule applies for more complicated binomials.

$$\left(-\frac{1}{2}x + \frac{5}{2}\right)^2 = \left(-\frac{1}{2}x\right)^2 + 2\left(-\frac{1}{2}x \cdot \frac{5}{2}\right) + \left(\frac{5}{2}\right)^2 = \frac{1}{4}x^2 - \frac{5}{2}x + \frac{25}{4}$$

When you multiply a difference of two monomials by their sum, like $(x+3)(x-3)$ or $(5x-2)(5x+2)$, the inner and outer products turn out to be opposites. Adding opposites gives you 0, so the middle term you would usually get disappears.

$$(x+3)(x-3) = x^2 - 3x + 3x - 3^2 = x^2 - 9$$

$$(5x-2)(5x+2) = (5x)^2 + 2 \cdot 5x - 2 \cdot 5x - 2^2 = 25x^2 - 4$$

Multiplying a sum of two monomials by their difference gives you the difference of the square of the first term and the square of the last term.

$$(a+b)^2 = a^2 + 2ab + b^2$$
$$(a-b)^2 = a^2 - 2ab + b^2$$
$$(a+b)(a-b) = a^2 - b^2$$

ALGEBRA TRAP

Don't forget to square the coefficient of the variable term when you square a binomial. $(3x+5)^2 = (3x+5)(3x+5)$, and when you FOIL, $3x \cdot 3x$ is $9x^2$, not $3x^2$.

CHECK POINT

Use special product shortcuts to complete each multiplication.

21. $(t-4)^2$

22. $(y-7)(y+7)$

23. $(2x-1)^2$

24. $\left(x+\dfrac{1}{2}\right)\left(x-\dfrac{1}{2}\right)$

25. $(2t-5)(2t+5)$

26. $(3x+5)^2$

27. $(5a-3)^2$

28. $(x^2-5)^2$

29. $(x^2+1)(x^2-1)$

30. $(5t^2-2t)(5t^2+2t)$

You already know some strategies for polynomial multiplication. The distributive property is your method for a monomial multiplied by a larger polynomial, and the FOIL rule applies to the specific but common situation of a binomial multiplied by a binomial. Now we need to have a look at how to deal with larger problems.

Larger Products

The general rule for multiplying two polynomials, of any size, is that you multiply each term of the first polynomial by each term of the second polynomial. For small problems, that's not too difficult to keep track of, but when the problems are larger, we need a way of keeping things organized.

Being asked to multiply $(2x^2+7x-3)(x^3-4x^2+8x+3)$ can be challenging if you look at it all at once. The trick is to break it into lots of little multiplications and then put the like terms together at the end. You could just multiply $2x^2$ times each term of the second polynomial, then multiply $7x$ times each term, and then multiply -3 times each, but it's easy to lose your place or miss a term. So a vertical arrangement will make it easier to stay organized. Place the polynomials one under the other, usually with the longer one on top.

$$x^3-4x^2+8x+3$$
$$\underline{2x^2+7x-3}$$

Start with the rightmost term of the bottom polynomial, the -3, and multiply each term of the upper polynomial by -3.

$$x^3-4x^2+8x+3$$
$$\underline{2x^2+7x-3}$$
$$-3x^3+12x^2-24x-9$$

Place a 0 under the rightmost term of that *partial product*. Then multiply each term of the upper polynomial by the $7x$.

$$
\begin{array}{r}
x^3 - 4x^2 + 8x + 3 \\
2x^2 + 7x - 3 \\
\hline
-3x^3 + 12x^2 - 24x - 9 \\
7x^4 - 28x^3 + 56x^2 + 21x + 0
\end{array}
$$

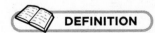 **DEFINITION**

> A **partial product** is the polynomial produced by multiplying by just one digit or term in a multiplication of polynomials or in the multiplication of multi-digit numbers.

For the third line, start with two zero terms, then multiply each term of the upper polynomial by $2x^2$.

$$
\begin{array}{r}
x^3 - 4x^2 + 8x + 3 \\
2x^2 + 7x - 3 \\
\hline
-3x^3 + 12x^2 - 24x - 9 \\
7x^4 - 28x^3 + 56x^2 + 21x + 0 \\
2x^5 - 8x^4 + 16x^3 + 6x^2 + 0 + 0
\end{array}
$$

Once you've multiplied by each term of the bottom polynomial, the last step is to combine like terms, and they should be nicely lined up in columns.

$$
\begin{array}{r}
x^3 - 4x^2 + 8x + 3 \\
2x^2 + 7x - 3 \\
\hline
-3x^3 + 12x^2 - 24x - 9 \\
7x^4 - 28x^3 + 56x^2 + 21x + 0 \\
2x^5 - 8x^4 + 16x^3 + 6x^2 + 0 + 0 \\
\hline
2x^5 - x^4 - 15x^3 + 74x^2 - 3x - 9
\end{array}
$$

If one or both polynomials have missing powers of the variable, you can insert zero terms to hold a place and help keep like terms aligned. The product $(t^2 - 3t + 2)(t^5 - 7t^3 + 5)$ can be rewritten as $(t^2 - 3t + 2)(t^5 + 0t^4 - 7t^3 + 0t^2 + 0t + 5)$. When you set up the vertical format, it looks like this.

$$t^5 + 0t^4 - 7t^3 + 0t^2 + 0t + 5$$
$$t^2 - 3t + 2$$

$$\begin{array}{r} 2t^5 + 0t^4 - 14t^3 + 0t^2 + 0t + 10 \\ -3t^6 + 0t^5 + 21t^4 + 0t^3 + 0t^2 - 15t + 0 \\ t^7 + 0t^6 - 7t^5 + 0t^4 + 0t^3 + 5t^2 + 0t + 0 \\ \hline t^7 - 3t^6 - 5t^5 + 21t^4 - 14t^3 + 5t^2 - 15t + 10 \end{array}$$

CHECK POINT

Complete each multiplication. Use the format that feels most comfortable.

31. $(4x - 3)(3x^2 + x - 6)$

32. $(x + 5)(x^2 + 2x + 3)$

33. $(2y - 3)(y^2 - 3y + 2)$

34. $(2a - 5)(a^3 - 5a^2 + a - 3)$

35. $(3x - 2)(x^2 - 3x - 6)$

36. $(2y - 3)(4y^3 - y^2 + 2y + 5)$

37. $(x - 2)(x^3 + 5x - 7)$

38. $(3a + 1)(a^2 - a)$

39. $(2x^3 - 1)(6x^4 + 3x^2 + 1)$

40. $(t - 4)(t^2 + 4t + 16)$

Dividing Polynomials

Once you know how to multiply polynomials, it's natural to think about the inverse, or opposite operation. You won't often need to divide polynomials, but there are two techniques you can use for this operation.

Single Term Divisor

Just as the basis of all polynomial multiplication is the multiplication of one monomial by another monomial, the root of all division of polynomials is division by a single term. Dividing a monomial by a monomial means that you divide the coefficients, and then deal with dividing the variables by again following the rules for dividing powers of the same base.

To divide $-8x^3$ by $4x$, think of the quotient $\dfrac{-8x^3}{4x}$ as $\dfrac{-8}{4}\cdot\dfrac{x^3}{x}$. The numerical coefficient will be -2, and $\dfrac{x^3}{x}=\dfrac{x^3}{x^1}=x^{3-1}=x^2$, so $\dfrac{-8x^3}{4x}=-2x^2$.

 TIP

If you're dividing a polynomial by a monomial and you encounter one or more terms for which the coefficients don't divide evenly, you'll find it more convenient to leave the quotients as improper fractions rather than decimals or mixed numbers. Simplify the fractions but don't fuss with decimals that might not terminate.

To divide a larger polynomial by a monomial, for example, $\dfrac{9x^3-12x^2+15x}{-6x}$, you'll break the quotient into a sum of divisions.

$$\frac{9x^3-12x^2+15x}{-6x}=\frac{9x^3}{-6x}+\frac{-12x^2}{-6x}+\frac{15x}{-6x}$$

Then do each division of monomial by monomial.

$$\frac{9x^3-12x^2+15x}{-6x}=\frac{9x^3}{-6x}+\frac{-12x^2}{-6x}+\frac{15x}{-6x}=-\frac{3}{2}x^2+2x-\frac{5}{2}$$

Not every division problem involving polynomials will produce a polynomial. Remember that the monomials that make up a polynomial must involve positive integer powers of the variable. The exponents can't be negative in a polynomial, but dividing powers could certainly produce negative exponents.

$$\frac{6y^5+8y^3-4y^2+2y-3}{2y^2}=\frac{6y^5}{2y^2}+\frac{8y^3}{2y^2}+\frac{-4y^2}{2y^2}+\frac{2y}{2y^2}+\frac{-3}{2y^2}=3y^3+4y-2+y^{-1}-\frac{3}{2}y^{-2}$$

You can give that quotient using negative exponents, with the understanding that the quotient is not a polynomial. You can use the language of your early experience with division and talk about a remainder. The first three terms of the quotient form a polynomial. You could take the division that far, and say that $6y^5+8y^3-4y^2+2y-3$ divided by $2y^2$ is $3y^3+4y-2$ with a remainder of $2y-3$. Or you could write the remainder over the divisor as a fraction and say $\dfrac{6y^5+8y^3-4y^2+2y-3}{2y^2}=3y^3+4y-2+\dfrac{2y-3}{2y^2}$.

CHECK POINT

Perform each division, putting your answer in simplest form.

41. $(8x + 24) \div 8$

42. $(28x^2 + 84x) \div 7x$

43. $(6x^5 - 12x^4 + 9x^3) \div 3x$

44. $(15y^3 - 20y^2 - 10y) \div -5y$

45. $(8x^4 - 12x^3) \div (4x^2)$

46. $(15t^6 - 21t^3) \div (-3t^2)$

47. $(16y^5 - 24y^3) \div (-8y^3)$

48. $(14x^6 - 42x^4 + 56x^2) \div (-14x^2)$

49. $(-9a^4 + 27a^3 - 81a^2) \div (-9a^2)$

50. $(24x^7 - 16x^5 - 48x^4 + 36x^3) \div (-4x^2)$

Long Division

When the divisor is a polynomial with more than one term, you can't use the term-by-term division strategy that you used with a monomial divisor. Instead, you have to fall back on the *algorithm* for long division that we learned in arithmetic.

DEFINITION

An **algorithm** is a step-by-step procedure for performing a task.

To divide $12x^3 + 8x^2 - 10x + 14$ by $x + 2$, set up a long division problem with $12x^3 + 8x^2 - 10x + 14$ as the dividend and $x + 2$ as the divisor.

$$x + 2 \overline{)12x^3 + 8x^2 - 10x + 14}$$

In arithmetic, you often used estimates to get you through long division. In polynomial long division, you can estimate using the first term of the divisor. In this case, that's x. Start by asking what $12x^3 \div x$ will give you. The answer to that is $12x^2$, so put $12x^2$ up in the quotient.

$$x + 2 \overline{)\begin{array}{l} 12x^2 \\ 12x^3 + 8x^2 - 10x + 14 \end{array}}$$

Of course, $12x^2$ isn't the whole quotient. It's just the first term. Multiply $12x^2(x + 2)$, to get $12x^3 + 24x^2$, and put that underneath the first two terms of the dividend.

$$x + 2 \overline{)\begin{array}{l} 12x^2 \\ 12x^3 + 8x^2 - 10x + 14 \\ \underline{12x^3 + 24x^2} \end{array}}$$

Subtract $12x^3 + 24x^2$ from the terms above, remembering that both terms are subtracted, not just the first one.

$$
\begin{array}{r}
12x^2 \\
x+2\overline{)12x^3 +8x^2 -10x +14} \\
\underline{12x^3 +24x^2} \\
0x^3 -16x^2
\end{array}
$$

The $0x^3$ can be dropped (the first term of your difference will always go away). Now, bring down the last two terms of the dividend.

$$
\begin{array}{r}
12x^2 \\
x+2\overline{)12x^3 +8x^2 -10x +14} \\
\underline{12x^3 +24x^2} \downarrow \quad \downarrow \\
-16x^2 -10x +14
\end{array}
$$

At this point, you have to decide if you're done. Will continuing produce negative exponents in the quotient? The quick way to decide is to compare $-16x^2 -10x +14$ to the divisor $x+2$. Because the degree of $-16x^2 -10x +14$ is higher than the degree of $x+2$, you can continue.

Continuing the division process means dividing $-16x^2 -10x +14$ by $x+2$. Follow the same steps. Estimate by dividing $-16x^2$ by x, then multiply that estimate times $x+2$, place that product underneath, subtract, and bring down any terms from the dividend not yet used. Here's how it looks.

$$
\begin{array}{r}
12x^2 -16x \\
x+2\overline{)12x^3 +8x^2 -10x +14} \\
\underline{12x^3 +24x^2} \downarrow \quad \downarrow \\
-16x^2 -10x +14 \\
\underline{-16x^2 -32x} \downarrow \\
22x +14
\end{array}
$$

Compare $22x+14$ to $x+2$. If the degree of the remainder polynomial is lower than the degree of the divisor, it's time to stop, but you're not there yet. $22x+14$ and $x+2$ have the same degree, so you can go one more round.

$$
\begin{array}{r}
12x^2 -16x +22 \\
x+2{\overline{\smash{\big)}\,12x^3 +8x^2 -10x +14}} \\
\underline{12x^3 +24x^2} \quad \downarrow \quad \downarrow \\
-16x^2 -10x +14 \\
\underline{-16x^2 -32x} \quad \downarrow \\
22x +14 \\
\underline{22x +44} \\
-30
\end{array}
$$

Once the remainder is an expression with a lower degree than the divisor, you can stop dividing. You can say that $12x^3 +8x^2 -10x+14$ divided by $x+2$ is $12x^2 -16x +22$ with a remainder of -30, or you can put that remainder over the divisor as a fraction and say the quotient is $12x^2 -16x +22-\dfrac{30}{x+2}$. You can check your work by multiplying the quotient times the divisor, then adding on the remainder. If the result is the original dividend, our work is correct.

$$
\begin{aligned}
(x+2)(12x^2 -16x+22)-30 &= \left(12x^3 -16x^2 +22x +24x^2 -32x +44\right)-30 \\
&= \left(12x^3 +8x^2 -10x +44\right)-30 \\
&= 12x^3 +8x^2 -10x +14
\end{aligned}
$$

If either the dividend or the divisor is missing powers of the variable, you'll probably find it easier to manage the division if you insert terms with zero coefficients to fill the spaces of the missing terms. If you need to divide $x^3 -8$ by $x-2$, set it up as $\left(x^3 +0x^2 +0x -8\right)\div(x-2)$.

$$
\begin{array}{r}
x^2 +2x +4 \\
x-2{\overline{\smash{\big)}\,x^3 +0x^2 +0x -8}} \\
\underline{x^3 -2x^2} \\
2x^2 +0x -8 \\
\underline{2x^2 -4x} \\
4x -8 \\
\underline{4x -8} \\
0
\end{array}
$$

CHECK POINT

Complete each long division and express the result as quotient plus remainder over divisor.

51. $\left(x^2 + 11x + 30\right) \div \left(x + 5\right)$

52. $\left(x^2 + 10x + 24\right) \div \left(x + 4\right)$

53. $\left(z^2 - 15z + 56\right) \div \left(z - 8\right)$

54. $\left(b^2 - 3b - 28\right) \div \left(b + 4\right)$

55. $\left(6x^2 + 13x + 6\right) \div \left(3x + 2\right)$

56. $\left(7y^2 - 3y - 4\right) \div \left(y - 1\right)$

57. $\left(9x^2 - 42x + 45\right) \div \left(3x - 8\right)$

58. $\left(2b^2 - 7b + 3\right) \div \left(b - 3\right)$

59. $\left(x^4 - 2x^2 + 1\right) \div \left(x^2 - 2x + 1\right)$

60. $\left(3x^4 - 17x^2 + 10\right) \div \left(x^2 - 5x\right)$

The Least You Need to Know

- To multiply a monomial times a polynomial, use the distributive property.
- To multiply two binomials, use the FOIL rule.
- For larger polynomials, arrange them vertically and multiply each term of the bottom polynomial by each term of the top one, and then combine like terms.
- To divide a polynomial by a monomial, divide each term of the polynomial by the monomial.
- To divide by a larger polynomial, use the long division algorithm.

Factoring

Throughout algebra, we talk a great deal about opposite operations. That makes sense, because inverse operations are the key to solving equations. We undo addition by subtraction and multiplication by dividing. That strategy is successful as long as we know what to subtract, or what the divisor should be.

Sometimes, however, we can see that an expression was probably created by multiplying, but we don't know either of the numbers that were multiplied. Without some number to use as the divisor, we can't really divide, so we're left to try to figure out what the numbers might have been.

If you're looking at the number 210 and you know that it was formed by multiplying 6 times some number, it's a simple matter to divide 210 by 6 and find that the other number is 35. On the other hand, if you're looking at 210 and you know it was created by multiplying two numbers but you don't know what either of the numbers was, you can only look for all the pairs of numbers that multiply to 210. You'll often find that there are several pairs of numbers. The number 210 could be $1 \cdot 210$, $2 \cdot 105$, $3 \cdot 70$, $5 \cdot 42$, $6 \cdot 35$, $7 \cdot 30$, $10 \cdot 21$, or $14 \cdot 15$.

In This Chapter

- Factoring out a greatest common factor
- Factoring trinomials by reversing the FOIL rule
- Factoring polynomials produced by special multiplication patterns

In this chapter, we'll investigate a similar search, but we'll work with polynomials rather than numbers. We'll look at a polynomial, think of it as a product and try to find two polynomials that will produce that product. There will be a lot of trial and error involved, so patience is required, but we'll also call on what we know about polynomial multiplication and division and on rules for exponents.

Greatest Common Factor

Your goal throughout this chapter (and in several that follow) will be to rewrite a polynomial as the product of two polynomials of lower degree. This is the process called *factoring*. The first type of polynomial multiplication you learned about was a monomial multiplied by a polynomial, and you did that by using the distributive property. The first type of factoring you'll look at is rewriting a polynomial as a monomial multiplied by a polynomial. You'll want to find out whether the polynomial you're trying to factor could have been created by multiplying using the distributive property, and if so, what the monomial was. Once you have that, you can divide to find the other polynomial.

> 📖 **DEFINITION**
>
> **Factoring** is the process of rewriting an integer as the product of two integers, or a polynomial as the product of two polynomials of lesser degree.

Let's start with something simple, or at least, short: $12x + 8$. Imagine a multiplication problem that uses the distributive law and that has $12x + 8$ as its answer.

$$\underline{\quad}(\underline{\quad} + \underline{\quad}) = 12x + 8.$$

You need to fill in the blanks, and like looking for the factors of 210, there could be more than one possibility. The largest possible polynomial that will fit into the monomial slot in this problem is the greatest common factor of the two terms $12x$ and 8. That means it's a monomial that divides both $12x$ and 8.

Like many other problems, this one can be broken down into smaller tasks. Look first at the coefficients and find the largest number that divides both, the largest factor they have in common. The largest number that divides both 12 and 8 is 4, because $12 = 4 \cdot 3$ and $8 = 4 \cdot 2$. Next consider the variable part of the terms. $12x$ contains x to the first power, but 8 is a constant, so we can either say it has no x or we can say it has x to the zero power. The greatest common factor will contain the smaller of the two powers of x, so it will have x^0. The greatest common factor of $12x$ and 8 is the 4 that is the greatest common factor of 12 and 8 multiplied by the x^0, the largest power of x that could divide both terms without creating negative exponents.

Now you know that the greatest common factor, or GCF, of $12x+8$ is $4x^0$, or simply 4, so the problem looks like this.

$$4(\underline{\quad} + \underline{\quad}) = 12x + 8$$

You can get the rest by dividing $\dfrac{12x+8}{4} = \dfrac{12x}{4} + \dfrac{8}{4} = 3x + 2$.

That means the multiplication problem that creates $12x+8$ is

$$4(\underline{3x} + \underline{2}) = 12x + 8$$

You can check our factoring by multiplying $4(\underline{3x} + \underline{2})$ and verifying that it equals $12x+8$.

TIP

You should find that with practice, you won't need to formally divide by the common factor. You actually thought about it already when you chose the common factor, and should just need to check your arithmetic.

Let's try one a little more challenging. Let's factor $15x^3 - 25x^2 + 10x$ and take it step by step.

First, what's the largest number that divides 15, -25, and 10? Because $15 = 5 \cdot 3$, $-25 = 5(-5)$, and $10 = 5 \cdot 2$, the GCF is 5.

Second, what's the largest power of x that will divide x^3, x^2, and x, without creating negative exponents? That would be the lowest power of the group, so x.

Third, identify the GCF of the polynomial. In this case, $5x$.

Fourth, divide the original polynomial by the GCF.

$$\frac{15x^3 - 25x^2 + 10x}{5x} = \frac{15x^3}{5x} - \frac{25x^2}{5x} + \frac{10x}{5x} = 3x^2 - 5x + 2$$

Finally, write the polynomial in factored form, as if it were a multiplication problem.

$$15x^3 - 25x^2 + 10x = 5x(3x^2 - 5x + 2)$$

TIP

If the common factor is positive, the signs of each term of the polynomial will stay the same. If the common factor is negative, all the signs of the polynomial will reverse. It's all or nothing.

Some people find it easier to factor by looking at a version of the polynomial written in an expanded form, like this.

$$15x^3 - 25x^2 + 10x = 5 \cdot 3 \cdot x \cdot x \cdot x - 5 \cdot 5 \cdot x \cdot x + 5 \cdot 2 \cdot x$$

In this expanded version, you can identify what appears in every term. Every term has a 5 and an x.

$$15x^3 - 25x^2 + 10x = \underline{5} \cdot 3 \cdot \underline{x} \cdot x \cdot x - \underline{5} \cdot 5 \cdot \underline{x} \cdot x + \underline{5} \cdot 2 \cdot \underline{x}$$

This identifies the common factor, which you can remove from the individual terms and place in front, once.

$$15x^3 - 25x^2 + 10x = \underline{5} \cdot 3 \cdot \underline{x} \cdot x \cdot x - \underline{5} \cdot 5 \cdot \underline{x} \cdot x + \underline{5} \cdot 2 \cdot \underline{x}$$
$$= 5x\left(\cancel{5} \cdot 3 \cdot \cancel{x} \cdot x \cdot x - \cancel{5} \cdot 5 \cdot \cancel{x} \cdot x + \cancel{5} \cdot 2 \cdot \cancel{x}\right)$$
$$= 5x\left(3 \cdot x \cdot x - \cdot 5 \cdot x + 2\right)$$
$$= 5x\left(3x^2 - 5x + 2\right)$$

Don't be surprised to see more than one variable in problems like these. When you first see a polynomial like $21ky^5 - 35ky^3 + 49ky^2$, the presence of two variables makes it look difficult. Once you identify the greatest common factor, however, the expression looks more manageable. All three coefficients are divisible by 7, all three terms have a single k, and each term contains at least y^2. The greatest common factor for $21ky^5 - 35ky^3 + 49ky^2$ is $7ky^2$. Divide $21ky^5 - 35ky^3 + 49ky^2$ by $7ky^2$ to find the other factor.

$$\frac{21ky^5 - 35ky^3 + 49ky^2}{7ky^2} = \frac{21ky^5}{7ky^2} - \frac{35ky^3}{7ky^2} + \frac{49ky^2}{7ky^2} = 3y^3 - 5y + 7$$

The polynomial $21ky^5 - 35ky^3 + 49ky^2$ can be factored as $7ky^2\left(3y^3 - 5y + 7\right)$. There are still two variables in the expression, but now we can see it as a monomial with two variables times a trinomial in y.

 TIP

Always look for a greatest common factor first. It may be the only type of factoring possible, or there may be other ways to factor the polynomial, but removing the GCF will make any other factoring easier to find.

If there is no way to factor a polynomial except to say that it equals itself times 1, the polynomial is prime, just as a number is prime when its only factors are itself and 1. The number 11 is prime because it can only be written as $11 \cdot 1$, and $x^2 - 7xy^2 + 5y^2$ is prime because there is no way to factor it except $1\left(x^2 - 7xy^2 + 5y^2\right)$.

CHECK POINT

Factor each expression by factoring out the greatest common factor. If no factoring is possible, state that the polynomial is prime.

1. $12x^2 - 18x$
2. $21t^2 - 35$
3. $16y^3 - 56y^2 + 72y$
4. $5ax^2 - 25ax + 15a$
5. $32x^5 - 48x^4 + 16x^3$
6. $11 - 77t + 22t^2$
7. $\pi r^2 + \pi rh$
8. $15x^2 - 3x - 5$
9. $y^2 + 2xy$
10. $4a^3 + 8a^2 - 24a$

The first type of polynomial multiplication we learned was the distributive property and the first type of factoring we explored was greatest common factor. The relationship between the two is clear. Factoring out the GCF is based on the assumption that the polynomial was created by doing a distributive multiplication.

The second type of multiplication we encountered was the FOIL rule, so it makes sense to look next at the factoring pattern that connects to FOIL. We're going to take this one in steps, starting with simple trinomials that fit closely to a pattern, and then branching out into others that require more trial and error.

Factoring $x^2 + bx + c$

When you multiply two simple binomials, like $(x+5)(x-3)$, the FOIL rule tells you to do four multiplications, but usually there are like terms to combine, and you end up with a trinomial.

$$(x+5)(x-3) = x^2 \underbrace{-3x + 5x}_{\text{like terms}} - 3 \cdot 5 = x^2 + 2x - 15$$

When you see a trinomial with an x^2-term, an x-term, and a constant term, it's reasonable to think that it might have come from that sort of multiplication.

The simpler the trinomial is, the easier it is to determine what its factors are. If the x^2 term is just x^2, you can be confident that each of the factors began with just a simple x. Then you can look at the constant term of the trinomial and make a list of factor pairs. The numbers that can multiply to this constant are possibilities for the constant terms of the binomials.

If you start out with $x^2 + 8x + 15$, the x^2 tells you the factors start with x's.

$$x^2 + 8x + 15 = (x + \underline{})(x + \underline{})$$

Then you can look at the 15, and realize that to get 15, you need to multiply either $1 \cdot 15$ or $3 \cdot 5$. That means you have two possibilities.

$$x^2 + 8x + 15 = (x + 1)(x + 15) \text{ or } x^2 + 8x + 15 = (x + 3)(x + 5)$$

How do we decide which is correct? The obvious answer is multiply. Multiply $(x + 1)(x + 15)$ and see if you get $x^2 + 8x + 15$. No? Try multiplying $(x + 3)(x + 5)$. Did that one give you $x^2 + 8x + 15$? It should. But having to multiply to find out which one works could be very time-consuming, especially if there are a lot of possible factor pairs. Noticing the pattern of what usually happens will lead to a shortcut. When you multiply two simple binomials of the form $(x + p)(x + q)$, the resulting trinomial starts with an x^2, ends with the product of the numbers in the p and q positions, and has a middle term, an x-term, whose coefficient is the sum of p and q.

$$(x + p)(x + q) = x^2 + px + qx + pq = x^2 + (p + q)x + pq$$

In the example, you know that both $1 \cdot 15$ and $3 \cdot 5$ multiply to 15, but only 3 and 5 will add to give you $8x$.

$$x^2 + 8x + 15 = (x \overset{3 \cdot 5 = 15}{+ 3)(x + 5)}_{3 + 5 = 8}$$

TIP

When you start to look for factors, focus on the absolute values of the numbers. If the constant term is positive, the factors of the constant will add to the middle term. If the constant is negative, the middle term is the difference between them. Once you've found them, you can place the signs. If the constant is positive, both factors get the sign of the middle term. If the constant is negative, one gets a plus and one a minus. Test to see which arrangement gives the correct middle term.

Here's another example. Let's try to factor $x^2 - x - 6$, assuming it came from the multiplication of two binomials of the form $(x + p)(x + q)$. Multiplying x times x for the First will give you x^2, so you just have to find the constants. To get a 6, you could use $1 \cdot 6$ or $2 \cdot 3$. You want the result to be −6, so you need to multiply a positive number by a negative number, but which number is positive and which is negative? Trying to get −6 rather than 6 actually increases our list of possibilities. You can use $-1 \cdot 6$, $1 \cdot (-6)$, $-2 \cdot 3$ or $2 \cdot (-3)$.

You need a pair of numbers that will multiply to −6 and also add to −1, the coefficient of the x-term. Only one pair of numbers, $2 \cdot (-3)$, meet both requirements. The correct factoring is $x^2 - x - 6 = (x + 2)(x - 3)$.

Step by step, here's how to tackle this sort of factoring problem. To factor a polynomial of the form $x^2 + bx + c$:

1. Set up two parentheses for the two binomial factors. $x^2 + bx + c = (x \quad)(x \quad)$ Start each with an x, so that $x \cdot x = x^2$.

2. List factor pairs for the absolute value of the constant term, $|c|$.

3. If c is a positive number, choose the factor pair that adds to $|b|$. $x^2 - 7x + 12 = (x \quad 3)(x \quad 4)$ because $3 \cdot 4 = 12$ and $3 + 4 = 7$.

4. If c is a negative number, choose the pair that subtracts to $|b|$. $x^2 + 2x - 15 = (x \quad 5)(x \quad 3)$ because $5 \cdot 3 = 15$ and $5 - 3 = 2$.

5. Put one of the factors in each parenthesis. $x^2 + bx + c = (x \quad p)(x \quad q)$

6. If c is positive, both of the factors get the sign of b. $x^2 - 7x + 12 = (x - 3)(x - 4)$ because the x-term is negative.

7. If c is negative, one factor will be positive and one negative. The larger factor gets the sign of b. $x^2 + 2x - 15 = (x + 5)(x - 3)$. The bigger factor of 5 gets the positive sign of $2x$ and the smaller factor gets the negative sign.

8. Multiply using the FOIL rule to check.

CHECK POINT

Factor each trinomial to the product of two binomials. If the trinomial cannot be factored, state that it is prime.

11. $x^2 + 7x + 12$

12. $y^2 - 6x + 8$

13. $t^2 + 9t + 20$

14. $y^2 - 2y + 1$

15. $x^2 + 12x + 27$

16. $t^2 - t - 12$

17. $x^2 + 2x - 15$

18. $y^2 - 8y - 20$

19. $t^2 + 10t - 11$

20. $x^2 - 12x - 28$

When the coefficient of the squared term is 1, the simple square term, x^2, clearly is equal to $x \cdot x$. When that coefficient is anything other than 1, however, you have to think more carefully about how to produce the square term, and also about the ramifications of that change for the rest of the factoring process.

Factoring $ax^2 + bx + c$

Let's look at this first from a multiplication point of view. If you multiply $(x-9)(x+1)$, the First multiplication gives you $x \cdot x = x^2$, Outer is $x \cdot 1 = x$, Inner is $-9 \cdot x = -9x$, and Last is $-9 \cdot 1 = -9$. The like terms combine to give you $(x-9)(x+1) = x^2 - 8x - 9$.

Now let's look at the change that occurs when you change one coefficient. If you change $(x-9)(x+1)$ to $(2x-9)(x+1)$, it causes a change in both the First multiplication and the Outer.

$$(2x-9)(x+1) = 2x^2 + 2x - 9x - 9 = 2x^2 - 7x - 9$$

Adding a coefficient in front of one x has changed both the x-squared term and the x-term. Keep that in mind when you're trying to factor a trinomial with a leading coefficient that is not 1. Notice that where you put the new coefficient also affects the x-term. $(2x-9)(x+1)$ is not the same product as $(x-9)(2x+1)$. As you've seen, $(2x-9)(x+1) = 2x^2 - 7x - 9$, but $(x-9)(2x+1) = 2x^2 + x - 18x - 9 = 2x^2 - 17x - 9$.

 TIP

If your middle term has a small coefficient, look for factors that are not very different. For a large middle term, you're more likely to need one large and one small factor. If your constant is 35, for example, try 1 and 35 if you need a large middle term, but 5 and 7 for a small one.

Let's try factoring $3x^2 + 8x - 35$. Our earlier plan of looking for a pair of factors of 35 that will add or subtract to 8 is not going to work when the coefficient of x^2 is not 1. We're going to have to depend much more on trial and error. We'll start by setting up the first term of each binomial. To produce a leading coefficient of 3, we'll need $3 \cdot 1$.

$$3x^2 + 8x - 35 = (3x \quad)(x \quad)$$

Next we look at the possible factor pairs for 35. We could use $1 \cdot 35$ or $5 \cdot 7$. Neither of these pairs add or subtract to 8, but that's okay. The 3 will affect the Outer product before the Inner and Outer combine to make 8. The constant term of 35 is actually -35, which means one factor will be positive and one negative, and that will make it look like the Inner and Outer products are subtracting to 8.

Now it's trial and error time, but you can use a little number sense as well. The first factor pair, 1 and 35, doesn't subtract to 8 on its own, and multiplying the 35 by 3 will only make things worse. $(3x\ 1)(x\ 35)$ has no chance of producing $3x^2 + 8x - 35$, and you can see that even before you put the signs in.

There are four possibilities.

$$(3x\ 1)(x\ 35)$$
$$(3x\ 35)(x\ 1)$$
$$(3x\ 5)(x\ 7)$$
$$(3x\ 7)(x\ 5)$$

You don't need to check the First or Last products, because you set these up based on those. You just need to look at the Outer and Inner and see if they'll give you the correct middle term.

	Outer	Inner	Sum	Difference	
$(3x\ 1)(x\ 35)$	$105x$	$1x$	$106x$	$104x$	The first possibility would be too large, but the second isn't much better.
$(3x\ 35)(x\ 1)$	$3x$	$35x$	$38x$	$32x$	
$(3x\ 5)(x\ 7)$	$21x$	$5x$	$26x$	$16x$	Better, but still too big.
$(3x\ 7)(x\ 5)$	$15x$	$7x$	$22x$	$8x$	You can get the +8x you need with this factor pair, if the signs are right.

Now that you've determined the numbers to need use, you have to position the signs. You need a plus and a minus to produce the -35 in $3x^2 + 8x - 35$, but here again our old rule about the larger number getting the sign of the middle term may not hold up. The leading coefficient can confuse the issue. Instead, look at the Outer and Inner products. The larger of those is going to get the sign of the middle term, in this case, the plus.

$$3x^2 + 8x - 35 = \underset{Outer:15x}{\overset{Inner:7x}{(3x\ 7)(x\ 5)}} = (3x - 7)(x + 5)$$

In the examples we've looked at so far, the coefficient of the squared term was a prime number. There was only one possibility of factors for that coefficient. When that coefficient is not prime, and there are several possibilities for factors of the lead term, there's more trial and error involved and it's important that you do it in an organized manner.

Let's factor $6x^2 + 13x + 5$. The lead term, $6x^2$, could be factored as $1x \cdot 6x$ or as $2x \cdot 3x$, but fortunately the constant term, 5, is prime, so you'll only have $1 \cdot 5$ to think about there. Work in pencil and have a good eraser handy, because this is a trial and error process. You're going to need to change things.

Choose a pair of factors for the lead term. Set up your parentheses.

$$6x^2 + 13x + 5 = (x \quad)(6x \quad)$$

Place the factors of the constant term.

$$6x^2 + 13x + 5 = (x \quad 1)(6x \quad 5)$$

Find the Inner and Outer products.

$$6x^2 + 13x + 5 = (x \overset{Inner\ 6x}{\quad 1)(6x} \quad 5)$$
$$\underset{Outer\ 5x}{}$$

Can the Inner and Outer add or subtract to the middle term? In this case, $6x$ and $5x$, won't combine to $13x$.

Switch the positions of the factors of the constants.

$$6x^2 + 13x + 5 = (x \quad 5)(6x \quad 1)$$

Try the Inner and Outer again.

$$6x^2 + 13x + 5 = (x \overset{Inner\ 30x}{\quad 5)(6x} \quad 1)$$
$$\underset{Outer\ 1x}{}$$

These still won't combine to our middle term of $13x$. If you had other possibilities for the factors of the constant term, you'd repeat this trial process for each factor pair. Don't forget to do the switch of position before you give up on a factor pair.

If you've tried all the possibilities for the constant term, it's tempting to think the trinomial is not factorable, but before you declare it a prime polynomial, you need to try that other pair of factors for the lead term. Start over, this time with $2x \cdot 3x$ for the lead.

$$6x^2 + 13x + 5 = (2x \quad)(3x \quad)$$

Place the factors of the constant and check the Inner and the Outer.

$$6x^2 + 13x + 5 = (2x \overset{Inner\ 3x}{\quad 1)(3x} \quad 5)$$
$$\underset{Outer\ 10x}{}$$

This will give you $13x$, so you just need to place the signs, but if this didn't work, you'd switch the 1 and the 5 and try again. In this case, the factoring is $6x^2 + 13x + 5 = (2x + 1)(3x + 5)$.

TIP

Make lists of the possibilities for factors of the lead term and factors of the constant term. Choose a pair of factors for the lead term and stay with it until you have tried every pair of factors for the constant, in both orders. Only then should you cross off that pair of lead factors and try the next one. Repeat all the pairs of factors for the constant. Check off each pair as you try it so you don't miss anything.

To factor $ax^2 + bx + c$:

1. List factors of the lead term and factors of the constant.

2. Place the first set of factors for the lead term and the first set of factors for the constant.

3. Check the Inner and Outer product and see if they can combine to form the middle term.

4. If not, switch the positions of the factors of the constant term and check the middle term again.

5. Repeat for each possible factor pair for the constant term.

6. If all these fail, move to the next pair of factors for the lead term, and retry the factor pairs for the constant.

7. Repeat until you find Inner and Outer products that will create the middle term.

8. Place the signs.

 CHECK POINT

Factor each trinomial to the product of two binomials. If the trinomial cannot be factored, state that the trinomial is prime.

21. $8x^2 + 2x - 1$ 26. $2x^2 - 15x + 7$

22. $3a^2 + 2a - 8$ 27. $3y^2 - 14y - 5$

23. $7y^2 + 20y - 3$ 28. $10t^2 - 23t + 12$

24. $6x^2 + 5x - 6$ 29. $4x^2 - 17x - 15$

25. $3t^2 - 11t - 15$ 30. $10x^2 + 7x - 12$

Factoring Special Forms

When we talked about multiplication of polynomials, we mentioned two special cases. Knowing that multiplying the sum and difference of the same two terms gives you a difference of squares, or that there is a pattern to the perfect square trinomial, can save you time in multiplying. That knowledge can also help you to factor.

Perfect Square Trinomial

If you can recognize a perfect square trinomial when you see one, you'll save time factoring. Remember $(x+b)^2 = x^2 + 2bx + b^2$ and $(ax+b)^2 = a^2 x^2 + 2abx + b^2$. When you look at $x^2 + 10x + 25$ and notice that both x^2 and 25 are perfect squares, look at the middle term. When you see that $10x = 2 \cdot 5x$, you can identify $x^2 + 10x + 25$ as a perfect square trinomial and know that it factors as $x^2 + 10x + 25 = (x+5)^2$.

If you're asked to factor $36x^2 - 84x + 49$, you might be reluctant to start trial and error, because of how many factor pairs exist for 36. Do you notice, however that 36 is 6^2? And that 49 is 7^2? Make a quick check of the middle term. Because $2 \cdot 6 \cdot 7 = 84$, this looks like a perfect square. $36x^2 - 84x + 49 = (6x - 7)^2$.

Difference of Squares

When you're confronted with a polynomial like $x^2 - 81$ or $100x^2 - 121$ to factor, you may feel helpless at first. How do you check the middle term if there isn't one? If you know the special cases from multiplication, however, you will recognize $x^2 - 81$ as $x^2 - 9^2$ and $100x^2 - 121$ as $10^2 x^2 - 11^2$, and realize that each of them will factor to the sum and difference of the same two terms.

$$x^2 - 81 = (x+9)(x-9) \text{ and } 100x^2 - 121 = (10x+11)(10x-11).$$

 ALGEBRA TRAP

When factoring a second degree expression with only two terms, it will be a difference of squares, or there will be a common factor, (or both), or it will be prime. For the difference of squares, both terms must be perfect squares. Non-squares will not work. It must be the difference of squares. The sum of squares does not factor.

CHECK POINT

Use your knowledge of special forms to factor each expression.

31. $x^2 - 64$

32. $y^2 + 12y + 36$

33. $4x^2 - 121$

34. $a^2 - 16a + 64$

35. $9y^2 - 49$

36. $25t^2 + 30t + 9$

37. $81t^2 - 16$

38. $16x^2 - 56x + 49$

39. $100x^2 - 1$

40. $36z^2 + 60z + 25$

The Least You Need to Know

- Always check for a common factor for all the terms of a polynomial. Remove it by writing the polynomial as a product for the distributive property.
- Factor $x^2 + bx + c = (x + p)(x + q)$, where $p \cdot q = c$ and $p + q = b$.
- You can factor $ax^2 + bx + c$ by making a list of the factors of a and a list of the factors of c. Try factor pairs in an organized fashion until you find the combination that produces your middle term.
- Remember special forms: $a^2x^2 + 2abx + b^2 = (ax + b)^2$ and $a^2x^2 - b^2 = (ax + b)(ax - b)$.

Radical, Quadratic, and Rational Functions

The last portion of one of my detective stories is always the most interesting part for me. The heroes have finished their investigation, overcome the obstacles, debunked the false leads, and now they're ready to unravel the mystery for us. They explain all the twists and turns, show us how this amazingly complicated problem breaks down to an application of things we already knew, and leave us thinking, "Of course! That makes so much sense!"

This part of the book is our chance to bring all the algebra you already know to explaining some problems that may seem mysterious at first. The graphs of square root functions, quadratic functions, and rational functions look very different from the lines we graphed earlier, but we'll be able to use many of the same strategies for graphing them. Solving equations related to these functions will also break down to simpler problems if we use techniques we learned earlier. It's time to tie all the pieces together.

Radical Functions

In arithmetic, we first learn to work with numbers and the four basic operations, but soon we introduce exponents, or powers, into the work, to shorten repeated multiplication. With exponents come roots, a way to undo a power. The symbol for a root, the radical sign, gives its name to the operation. The square root of 2, written $\sqrt{2}$, is often read as "radical two."

Roots of numbers are numbers, and you've already learned a bit about how to simplify them. When variables start to appear under the radical sign, simplifying becomes a little more difficult but even more important. When variable expressions under radicals appear in equations, the tactics you've used for linear equations are no longer enough. In this chapter, we'll look at the process of simplifying radicals that contain variables, and we'll develop some new tools for solving radical equations.

In This Chapter

- Simplifying radicals
- Rationalizing denominators
- Doing arithmetic with radicals
- Solving radical equations

Simplifying Radicals

Exponents tell how many times to use a number in a multiplication. If you write 3^5, you're saying you want to multiply five 3s. $3^5 = 3 \cdot 3 \cdot 3 \cdot 3 \cdot 3 = 243$. You can raise a number to any power you need, but the power you'll most commonly see is the second power, or square. Multiplying $9 \cdot 9$ can be written as 9^2. Formally, that's read as "9 to the second power," but casually we say "9 squared."

THINK ABOUT IT

Raising a number to the second power came to be known as squaring because finding the area of a square requires multiplying the length of a side by itself. You raise the length of a side to the second power, or square the side, to find the area of a square. To find the volume of a cube, you raise the length of an edge of the cube to the third power, which is why the third power is called the cube and raising a number to the third power is cubing.

Roots are the inverses, or opposites, of powers, and just as you can raise a number to any power, you can take any root. You know $3^5 = 243$, so the fifth root of 243 is 5. The fifth root tells you what number you would raise to the fifth power to produce 243. The symbol for the fifth root of 243 is $\sqrt[5]{243}$. The small number 5 in the crook of the radical sign is called the *index*. It tells what power is being undone. If you don't see an index, it's 2. The square root, the inverse of the second power is the most common, so if you don't see an index, you can assume it's a square root. The number under the radical sign is called the radicand.

DEFINITION

The **index** is the small number that appears in the crook of the radical sign and tells what power is being undone. If no index appears, the radical indicates a square root.

Working with radicals is easier if they are in simplest radical form. Getting to that form takes a few steps. The first step is to get the smallest possible number under the radical. Use the rule that $\sqrt{a} \cdot \sqrt{b} = \sqrt{a \cdot b}$. Think about numbers whose square root you know. $\sqrt{4} = 2$ and $\sqrt{9} = 3$. If you multiply $\sqrt{4} \cdot \sqrt{9}$, you're multiplying $2 \cdot 3$, which is 6, and 6 is $\sqrt{36}$, which is $\sqrt{4 \cdot 9}$.

TIP

The method of simplifying radicals works for all indices, not just square roots, but to simplify a third root, for example, you must look for numbers that are perfect cubes, not perfect squares. If you were simplifying a fifth root, you'd look for factors that were integers raised to the fifth power.

To get to the smallest number under the radical, look for factors of the radicand whose square root you know. To simplify $\sqrt{48}$, list the possible factor pairs for 48.

$$48 = 1 \cdot 48$$
$$= 2 \cdot 24$$
$$= 3 \cdot 16$$
$$= 4 \cdot 12$$
$$= 6 \cdot 8$$

There are two numbers on that list that are perfect squares, 4 and 16. Use the larger one to write $\sqrt{48}$ as $\sqrt{3 \cdot 16}$ and then use the rule to say $\sqrt{3 \cdot 16} = \sqrt{3} \cdot \sqrt{16}$. Because you know that $\sqrt{16} = 4$, you can say $\sqrt{48} = \sqrt{3 \cdot 16} = \sqrt{3} \cdot \sqrt{16} = 4\sqrt{3}$. The simplest form of $\sqrt{48}$ is $4\sqrt{3}$.

To put a radical in simplest form, find the largest perfect square that is a factor of the radicand. Rewrite as a product of two radicals and take the square root of the perfect square.

 TIP

To leave the smallest possible number under the radical, look for the largest perfect square factor. The fastest way to find the largest perfect square, however, is to start with a small factor and see if the factor-pair partner is a perfect square. If you need to simplify $\sqrt{432}$, don't grope for perfect square factors. Think $432 \div 2 = 216$, not a perfect square, but $432 \div 3 = 144$ and 144 is the largest perfect square factor.

When variables appear in the radicand, it becomes a little more challenging to simplify, because you don't know what the variable represents. $\sqrt{x}$ can't be simplified any further. The number that x represents might be a perfect square or might be a number with a perfect square factor, but you just don't know that. There are steps you can take, however.

If the variable has a numeric coefficient, you can simplify that. $\sqrt{4x} = \sqrt{4} \cdot \sqrt{x} = 2\sqrt{x}$ and $\sqrt{48x} = \sqrt{16 \cdot 3 \cdot x} = \sqrt{16} \cdot \sqrt{3} \cdot \sqrt{x} = 4\sqrt{3x}$. If the variable under the radical is raised to a power, you may be able to work with that to make the radicand smaller.

Roots undo powers, so taking the square root of a square should get you back where you started, and it does, with one caution. When you take a square root, you probably think of a positive number. In fact, it is agreed a radical in print will mean the principal, or positive, square root. $\sqrt{9}$ will mean 3, but we know that $(-3)^2$ is also equal to 9. By agreement, if we want to talk about the negative square root, we put a negative sign in front of the radical. $-\sqrt{9} = -3$. If we want both the positive and the negative square roots, we write $\pm\sqrt{9} = \pm3$.

When it comes time to take the square root of x^2, as in $\sqrt{x^2}$, the quick response is x. The problem is, we don't know what number x stands for. Suppose x were -5. $(-5)^2 = 25$ and $\sqrt{(-5)^2} = \sqrt{25} = 5$ but x is not 5. It's -5. In that case (and many others), saying $\sqrt{x^2} = x$ wouldn't be quite true.

$\sqrt{x^2}$ might equal x, or it might equal the opposite of x, whichever is the positive number. To guard against this problem, we will say $\sqrt{x^2} = |x|$.

THINK ABOUT IT

If you have a guarantee that x is positive, you can drop the absolute value signs. Sometimes that guarantee will be in the directions (Assume all variables are positive). Sometimes the situation may tell you negative numbers aren't reasonable, for example, when the variables represent lengths of sides.

Now you can apply a strategy for simplifying radicals to radicands that are variable terms. To simplify $\sqrt{72x^3}$, factor the radicand, using perfect squares as factors whenever you can.

$$\sqrt{72x^3} = \sqrt{36 \cdot 2 \cdot x^2 \cdot x}$$

Apply the rule.

$$\begin{aligned} \sqrt{72x^3} &= \sqrt{36 \cdot 2 \cdot x^2 \cdot x} \\ &= \sqrt{36} \cdot \sqrt{2} \cdot \sqrt{x^2} \cdot \sqrt{x} \\ &= 6 \cdot \sqrt{2} \cdot |x| \cdot \sqrt{x} \end{aligned}$$

Rearrange a bit, and apply the product rule in the other direction to get back to just one radical.

$$\begin{aligned} \sqrt{72x^3} &= \sqrt{36 \cdot 2 \cdot x^2 \cdot x} \\ &= 6|x|\sqrt{2}\sqrt{x} \\ &= 6|x|\sqrt{2x} \end{aligned}$$

ALGEBRA TRAP

There is no rule for the root of a sum except to find the sum and take its square root. $\sqrt{x+y}$ does *not* equal $\sqrt{x} + \sqrt{y}$. Putting numbers in place of x and y will show you that. $\sqrt{9+16} = \sqrt{25} = 5$ but $\sqrt{9} + \sqrt{16} = 3 + 4 = 7$. Clearly, they're not equivalent.

CHECK POINT

Put each radical in simplest form.

1. $\sqrt{8}$

2. $\sqrt{27}$

3. $\sqrt{128}$

4. $\sqrt{5x^2}$

5. $\sqrt{50x^2}$

6. $5\sqrt{32x^3}$

7. $\sqrt{98t^4}$

8. $\sqrt{x^3y^2}$

9. $9\sqrt{4a^3b^5}$

10. $5x\sqrt{12x^3}$

Making the radicand as small as possible is one step in simplifying a radical expression, and the one used most often. If the expression under the radical is a fraction or if the radical expression is in the denominator of a fraction, however, you have more work to do.

Rationalizing Denominators

The basic rule used to simplify radicals is $\sqrt{a} \cdot \sqrt{b} = \sqrt{a \cdot b}$. When dealing with fractions and radicals at the same time, we'll count on a related rule: $\sqrt{\dfrac{a}{b}} = \dfrac{\sqrt{a}}{\sqrt{b}}$. If we need to take the square root of $\dfrac{4}{9}$, this rule states that $\sqrt{\dfrac{4}{9}} = \dfrac{\sqrt{4}}{\sqrt{9}} = \dfrac{2}{3}$. It also means that we can go the other way. $\dfrac{\sqrt{48}}{\sqrt{8}} = \sqrt{\dfrac{48}{8}} = \sqrt{6}$.

Sometimes, however, things are not that tidy. If you need to take the square root of $\dfrac{1}{2}$, you can say $\sqrt{\dfrac{1}{2}} = \dfrac{\sqrt{1}}{\sqrt{2}}$ and you know that the square root of 1 is 1, but the square root of 2 is an irrational number. You don't want a non-terminating, non-repeating decimal in the denominator, and if you start rounding, you won't have an exact answer. So it seems like you're stuck with $\sqrt{\dfrac{1}{2}} = \dfrac{1}{\sqrt{2}}$.

One of the requirements for simplest radical form is that there are no radicals in the denominator. Radicals in the numerator are okay, but not the denominator. How can you get $\dfrac{1}{\sqrt{2}}$ to simplest radical form? By using a familiar method to change the way a fraction looks without changing its value: multiply by 1.

THINK ABOUT IT

Why are we willing to have radicals in the numerator but not in the denominator? The answer has to do with being able to have a sense of numbers without having to carry a calculator around constantly. If I write $\frac{\sqrt{3}}{2}$, you can think "$\sqrt{3}$ is about 1.7, so $\frac{\sqrt{3}}{2}$ is about 0.85." It's not exact, but you have a sense of the number. If, on the other hand, I write $\frac{2}{\sqrt{3}}$, even the estimated value of $\sqrt{3}$ as 1.7 doesn't help all that much, because you probably can't divide 2 by 1.7 in your head. It's more than 1, less than 2, but that's about the best we can do. But if we rationalize the denominator, $\frac{2}{\sqrt{3}} = \frac{2\sqrt{3}}{3} \approx \frac{2(1.7)}{3} \approx \frac{3.4}{3} \approx 1.13$.

To create a common denominator, multiply the numerator and denominator of the fraction by the same number. If you multiply $\frac{1}{3} \cdot \frac{7}{7} = \frac{7}{21}$, you create a very different looking fraction, $\frac{7}{21}$, that has the same value as $\frac{1}{3}$. The reason that the value doesn't change is that the multiplier, $\frac{7}{7}$, is equal to 1. Let's use a similar multiplier to get rid of the radical in the denominator of $\frac{1}{\sqrt{2}}$. Multiply by $\frac{\sqrt{2}}{\sqrt{2}}$, which is equal to 1.

$$\frac{1}{\sqrt{2}} \cdot \frac{\sqrt{2}}{\sqrt{2}} = \frac{\sqrt{2}}{\sqrt{2} \cdot \sqrt{2}} = \frac{\sqrt{2}}{\sqrt{4}} = \frac{\sqrt{2}}{2}$$

Clearing the radicals from the denominator of a fraction (even if it means putting radicals in the numerator) is called *rationalizing the denominator*. If the denominator of a fraction contains a single radical term, you can rationalize the denominator by multiplying the numerator and denominator by the radical of the denominator. It's not necessary to include any multiplier outside the radical. You can rationalize the denominator of $\frac{4\sqrt{2}}{5\sqrt{3}}$ by multiplying by $\frac{\sqrt{3}}{\sqrt{3}}$. You don't need to use the 5.

$$\frac{4\sqrt{2}}{5\sqrt{3}} = \frac{4\sqrt{2}}{5\sqrt{3}} \cdot \frac{\sqrt{3}}{\sqrt{3}} = \frac{4\sqrt{6}}{5\sqrt{9}} = \frac{4\sqrt{6}}{5 \cdot 3} = \frac{4\sqrt{6}}{15}$$

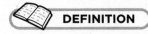

DEFINITION

Rationalizing the denominator is the process of changing the appearance, but not the value, of a quotient so that no radicals remain in the denominator.

It is possible for the denominator of the fraction to be a sum or difference of two terms, one or both of which is a radical. In that situation, the strategy you just used won't work. If you multiply $\frac{6}{5+\sqrt{2}}$ by $\frac{\sqrt{2}}{\sqrt{2}}$, the radical in the denominator moves, but it doesn't disappear.

$$\frac{6}{5+\sqrt{2}}\cdot\frac{\sqrt{2}}{\sqrt{2}}=\frac{6\sqrt{2}}{5\sqrt{2}+\sqrt{2}\cdot\sqrt{2}}=\frac{6\sqrt{2}}{5\sqrt{2}+2}$$

You haven't rationalized the denominator. If fact, things have gotten a little worse. When the denominator is a sum or difference, you will need to use the *conjugate* of the denominator. The conjugate is the same two terms with the opposite sign connecting them. The conjugate of $5+\sqrt{2}$ is $5-\sqrt{2}$ and the conjugate of $3-\sqrt{5}$ is $3+\sqrt{5}$.

DEFINITION

The **conjugate** of an expression that is the sum of two terms is an expression that is the difference of those two terms. If $a-b$ is the conjugate of $a+b$, then $a+b$ is the conjugate of $a-b$, and the product $(a-b)(a+b)=a^2-b^2$.

You need to multiply $\frac{6}{5+\sqrt{2}}$ by $\frac{5-\sqrt{2}}{5-\sqrt{2}}$. To do that multiplication, you need the distributive property and the FOIL rule covered in Chapter 2.

$$\frac{6}{5+\sqrt{2}}\cdot\frac{5-\sqrt{2}}{5-\sqrt{2}}=\frac{6\left(5-\sqrt{2}\right)}{\left(5+\sqrt{2}\right)\left(5-\sqrt{2}\right)}$$

$$=\frac{30-6\sqrt{2}}{25-5\sqrt{2}+5\sqrt{2}-\sqrt{2}\cdot\sqrt{2}}$$

$$=\frac{30-6\sqrt{2}}{25-2}$$

$$=\frac{30-6\sqrt{2}}{23}$$

ALGEBRA TRAP

Once you start canceling, it can be hard to stop. If you have $\frac{15+\sqrt{2}}{5}$, it's tempting to think you could cancel the 15 with the 5, but that wouldn't be correct. You can cancel a factor from the numerator with a factor from the denominator, but the 15 is a term, not a factor. It's connected to $\sqrt{2}$ by addition. Remember the fraction bar acts like a grouping symbol. You need to divide 15 by 5 and $\sqrt{2}$ by 5. $\frac{15+\sqrt{2}}{5}=\frac{15}{5}+\frac{\sqrt{2}}{5}$, not $3+\sqrt{2}$. Never cancel terms.

CHECK POINT

Rationalize denominators and put each expression in simplest radical form.

11. $\dfrac{8}{\sqrt{2x}}$

14. $\dfrac{5\sqrt{2}}{4\sqrt{10}}$

17. $\dfrac{5}{2+\sqrt{3}}$

12. $\dfrac{9}{2\sqrt{12}}$

15. $\dfrac{10}{\sqrt{50}}$

18. $\dfrac{1}{\sqrt{3}-8}$

13. $\dfrac{27}{4\sqrt{18}}$

16. $\dfrac{1}{3-\sqrt{3}}$

19. $\dfrac{2}{3-\sqrt{x}}$

20. $\dfrac{10}{\sqrt{7}-\sqrt{2}}$

Operations with Radicals

You've already done some multiplication with variables, and there really isn't any division. If you have a division problem involving radicals, you just write it as a fraction and put it in simplest form, being sure to rationalize the denominator.

When it comes to adding and subtracting, radicals act a lot like variables. You can only add or subtract like radicals, just as you can only combine like variable terms. You perform the operations by adding or subtracting the coefficients in front of the radicals. $6\sqrt{3}+9\sqrt{3}=15\sqrt{3}$ just as $6x+9x=15x$.

If you're trying to combine variable terms and they don't have the same variable, there's nothing you can do. $5x+3y$ just can't be simplified. When you're working with radicals, however, you don't want to jump to conclusions. At first glance, it may seem that $\sqrt{72}+\sqrt{98}$ is impossible to simplify, but each of those radicals can be simplified.

$$\begin{aligned}
\sqrt{72}+\sqrt{98} &= \sqrt{36\cdot 2}+\sqrt{49\cdot 2} \\
&= \sqrt{36}\cdot\sqrt{2}+\sqrt{49}\cdot\sqrt{2} \\
&= 6\sqrt{2}+7\sqrt{2} \\
&= 13\sqrt{2}
\end{aligned}$$

In their original form, the radicals appeared to be unlike, but once they're simplified, you can see that both are multiples of $\sqrt{2}$, and can be combined.

CHECK POINT

Simplify each expression by performing the indicated operations. Simplify radicals when necessary.

21. $\sqrt{8} - \sqrt{2}$

22. $\sqrt{75} + \sqrt{48}$

23. $3\sqrt{72} - \sqrt{98}$

24. $\sqrt{25a} + \sqrt{49a}$

25. $\sqrt{2x^2} - \sqrt{50x^2}$

26. $\dfrac{10}{\sqrt{5}} + 2\sqrt{45}$

27. $\dfrac{\sqrt{32}}{8} + \dfrac{2}{\sqrt{8}}$

28. $x\sqrt{2x} + \sqrt{18x^3}$, $x > 0$

29. $8\sqrt{24} + \dfrac{\sqrt{54}}{3} - 2\sqrt{96}$

30. $6\sqrt{ab^2} + 3b\sqrt{a} - b\sqrt{25a}$, $a > 0$, $b > 0$

Solving Radical Equations

When you encounter radicals in equations, you may need to add a new technique to your toolbox for solving equations. If the radical is just the square root of a constant, you can treat it like any other number. When the radicand includes a variable, on the other hand, you need a new way to get to a solution.

Equations with One Radical

If you're facing a linear equation like $6x - 7 = 11$, you can use inverse operations to isolate x.

$$6x - 7 = 11$$
$$6x - 7 + 7 = 11 + 7$$
$$6x = 18$$
$$\frac{6x}{6} = \frac{18}{6}$$
$$x = 3$$

When you meet a radical equation like $\sqrt{6x - 7} = 11$, you need to free the $6x - 7$ from that radical, so that you can get to work isolating x. To do that, use an inverse operation. The inverse of taking a square root is squaring, so begin by squaring both sides.

$$\sqrt{6x - 7} = 11$$
$$\left(\sqrt{6x - 7}\right)^2 = (11)^2$$
$$6x - 7 = 121$$

Squaring a radical makes the radical "lift" or disappear, and leaves you with just the radicand. Squaring the 11 on the other side gives us 121. The important part is that you now have a linear equation that you can solve as usual.

$$\sqrt{6x-7} = 11$$
$$\left(\sqrt{6x-7}\right)^2 = (11)^2$$
$$6x - 7 = 121$$
$$6x = 128$$
$$x = 21\frac{1}{3}$$

THINK ABOUT IT

Why does squaring both sides make the radical disappear? Squaring and taking the square root are opposite or inverse operations. Each will undo the work of the other. If you take the square root of a number ($\sqrt{49} = 7$) and then you square that answer ($7^2 = 49$) you get right back to the number you started with. When you square a square root, you get the radicand. The radical lifts.

When you solve a radical equation, the first important additional step is squaring both sides to remove the radical. There are a couple of other concerns that are important, too. One is to be certain to check the solution. The squaring of both sides that's critical to the whole process can also introduce what are called *extraneous solutions*. They're not mistakes. You've done your algebra correctly, but they don't work. Sometimes you get a perfectly normal solution; sometimes you get an extraneous solution. Later, when we look at equations that have more than one solution, you'll see that you might get two usable solutions, two extraneous solutions, or one of each. Whenever a radical is involved, it's crucial to check the solution.

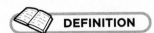

DEFINITION

An **extraneous solution** is a solution produced by correct algebraic procedures that does not satisfy the equation.

So let's check the solution of $x = 21\frac{1}{3}$ for the equation $\sqrt{6x-7} = 11$. Replace x in the original equation with $21\frac{1}{3}$, and simplify.

$$\sqrt{6x-7}=11$$

$$\sqrt{6\left(21\frac{1}{3}\right)-7}=11$$

$$\sqrt{126\frac{6}{3}-7}=11$$

$$\sqrt{128-7}=11$$

$$\sqrt{121}=11$$

$$11=11$$

The solution checks.

The other thing to keep in mind when you're solving radical equations is the radical must be isolated on one side before you square both sides. Suppose that, instead of $\sqrt{6x-7}=11$, you needed to solve $\sqrt{6x}-7=11$. Only the $6x$ is under the radical. If you tried to square both sides right now, you would have a problem, a couple of problems actually.

Problem 1 is that to square $\sqrt{6x}-7$, you're going to have to use the FOIL rule. $\left(\sqrt{6x}-7\right)^2=\left(\sqrt{6x}-7\right)\left(\sqrt{6x}-7\right)$ and that's a lot of work. Take a look at what happens if you put in that work.

$$\left(\sqrt{6x}-7\right)^2=\left(\sqrt{6x}-7\right)\left(\sqrt{6x}-7\right)$$
$$=\sqrt{6x}\cdot\sqrt{6x}-7\cdot\sqrt{6x}-7\cdot\sqrt{6x}+49$$
$$=6x-14\sqrt{6x}+49$$

Not only have you not eliminated the radical, you've produced something even more difficult to solve than the equation you started with. This is not a path you want to take.

Instead, take a moment to isolate the radical. In this equation, add 7 to both sides.

$$\sqrt{6x}-7=11$$
$$\sqrt{6x}=18$$

Now you can square both sides and the radical will lift.

$$\sqrt{6x}-7=11$$
$$\sqrt{6x}=18$$
$$\left(\sqrt{6x}\right)^2=(18)^2$$
$$6x=324$$
$$x=54$$

That's easier, right? Check the solution before moving on. $\sqrt{6 \cdot 54} - 7 = \sqrt{324} - 7 = 18 - 7 = 11$.

To solve an equation with one radical:

1. Isolate the radical.

2. Square both sides.

3. Solve the resulting equation.

4. Check your solution.

Equations with Two Radicals

Solving radical equations depends on eliminating the radicals, so it's natural that an equation that has more than one radical would look challenging. In fact, they'll yield to the method we've outlined. It just requires patience and persistence.

If you want to solve $\sqrt{3x+1} - \sqrt{x+11} = 0$, focus on isolating the first radical. That will give you $\sqrt{3x+1} = \sqrt{x+11}$, and squaring both sides will lift both radicals.

$$\sqrt{3x+1} = \sqrt{x+11}$$
$$\left(\sqrt{3x+1}\right)^2 = \left(\sqrt{x+11}\right)^2$$
$$3x+1 = x+11$$
$$2x+1 = 11$$
$$2x = 10$$
$$x = 5$$

The solution of $x = 5$ does check, and that wasn't really any more work than the equations with one radical. The key to that simplicity was the zero in the original equation. Let's look at one that's not quite so simple.

To solve the equation $\sqrt{x+1} - \sqrt{x-2} = 1$, start the same way, by isolating one radical. Then square both sides.

$$\sqrt{x+1} - \sqrt{x-2} = 1$$
$$\sqrt{x+1} = 1 + \sqrt{x-2}$$
$$\left(\sqrt{x+1}\right)^2 = \left(1+\sqrt{x-2}\right)^2$$

This is going to require the FOIL rule.

$$\left(\sqrt{x+1}\right)^2 = \left(1+\sqrt{x-2}\right)^2$$
$$x+1 = 1+2\sqrt{x-2}+x-2$$
$$1 = 1+2\sqrt{x-2}-2$$
$$1 = -1+2\sqrt{x-2}$$

Now isolate the other radical. Remember: patience and persistence.

$$1 = -1+2\sqrt{x-2}$$
$$2 = 2\sqrt{x-2}$$
$$1 = \sqrt{x-2}$$

Square both sides yet again, and solve.

$$1 = \sqrt{x-2}$$
$$1 = x-2$$
$$x = 3$$

Extraneous solutions often occur when there are variables on both sides of the equation. Here's an example of a radical equation that has two solutions, but one is extraneous. The solution below will be explained in a later chapter, but the important piece is that there are two solutions.

$$\sqrt{x-1} = x-7$$
$$x-1 = (x-7)^2$$
$$x-1 = x^2-14x+49$$
$$x^2-15x+50 = 0$$
$$(x-10)(x-5) = 0$$
$$x-10 = 0 \qquad x-5 = 0$$
$$x = 10 \qquad x = 5$$

The solution of $x=10$ checks: $\sqrt{10-1} = \sqrt{9} = 3 = 10-7$. The solution of $x=5$ is extraneous. $\sqrt{5-1} = \sqrt{4} = 2$ but $5-7 = -2$, not 2.

CHECK POINT

Solve each equation. Be certain to check for extraneous solutions.

31. $\sqrt{2x-7} = 3$

32. $\sqrt{x-1} = 4$

33. $\sqrt{x-3} - 2 = 5$

34. $\sqrt{2x-3} = 5$

35. $\sqrt{5x-1} - 11 = 3$

36. $3\sqrt{x+1} = 5$

37. $2\sqrt{x+2} - 3 = 0$

38. $\sqrt{x^2 - 11} = x - 1$

39. $\sqrt{\dfrac{2x+1}{3}} + 5 = 8$

40. $4\sqrt{x} = \dfrac{x+6}{\sqrt{x}}$

The Least You Need to Know

- Simplify radical expressions by finding the largest perfect square factor of the radicand, and apply the $\sqrt{ab} = \sqrt{a}\sqrt{b}$ rule.

- Rationalize a denominator that is a single term by multiplying the numerator and denominator by the radical that appears in the denominator.

- The conjugate of an expression of the form $a+b$ is an expression of the form $a-b$.

- Rationalize a two-term denominator by multiplying the numerator and denominator by the conjugate of the denominator.

- Solve radical equations by isolating the radical and squaring both sides.

- Always check for extraneous solutions when solving radical equations.

Quadratic Equations

You may have noticed as you worked your way through the preceding chapters that while we talked about rules for exponents, and expressions involving exponents, we haven't solved equations that involved exponents. Well, the time has come.

In this chapter, we'll look at methods for solving equations that involve a second power of the variable. Later in your exploration of algebra, you'll investigate other powers of the variable, but for now, we'll focus on equations involving variables squared. We'll call on many of the skills you've already learned, putting them together in new ways to solve equations from this new category.

Quadratic Equations

A *quadratic equation* is defined as an equation that is in the form (or can be put in the form) $ax^2 + bx + c = 0$, where a, b, and c are real numbers and $a \neq 0$. That means it's an equation that contains a term in which the variable is squared. There may be another variable term with the variable to the first power. There may be a constant term. Those terms may or may not be there, but what makes it a quadratic equation is the squared term.

In This Chapter

- Solving quadratic equations by taking square roots
- Using the quadratic formula to solve any quadratic equation
- Determining the number of solutions by looking at the discriminant
- Applying the Zero Product Property to solve a quadratic equation by factoring

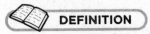 **DEFINITION**

> A **quadratic equation** is an equation that includes a term in which the variable is squared. There may also be a variable term in which the variable appears but is not squared, and a constant term.

Examples of quadratic equations include equations as simple as $x^2 = 4$, which fits the pattern if you think of it as $1x^2 + 0x - 4 = 0$. The equation $x(x+8) = 0$ is also an example of a quadratic equation, because although you don't see a square right away, when you distribute to remove the parentheses, it becomes $x^2 + 8x = 0$, a quadratic with a constant term of 0. Of course, many quadratics will look just like the model in the definition, like $3x^2 + 5x + 9 = 0$.

A key characteristic of quadratic equations is that quadratics have two solutions. If $x^2 = 4$, x might be 2 or x might be -2. Rules for signs state that both 2^2 and $(-2)^2$ equal 4. The two solutions of a quadratic equation won't always be a number and its opposite, but you can expect to see two solutions in most quadratics.

The Square Root Method

When you're faced with a simple quadratic equation, like $x^2 = 4$, you may be able to use a little logic and your knowledge of exponents to figure out the solutions. Asking yourself, "what number, times itself, gives me 4?" may be enough. You might need a little prodding to remember the negative possibility, but if you know to expect two solutions, that should work out.

What happens if you change that equation to $x^2 = 5$? You probably can't pull the numbers out of your head, because no integer, squared, gives you 5, but you know you want to ask that question: what number, squared, equals 5? The mathematical way to ask that question is, "what is the square root of 5?" If $x^2 = 5$, $x = \pm\sqrt{5}$. The square root of 5 is an irrational number, so you want to leave that answer in simplest radical form for an exact answer, or use a calculator (or table of square roots) to get an approximate answer.

This is the heart of the square root method. You isolate the squared variable, then take the square root of both sides, and put the answer in simplest radical form. Here's another example.

$$x^2 - 43 = 5$$

Isolate the x^2.

$$x^2 = 48$$

Take the square root of both sides. Don't forget, you want both the positive and the negative square root.

$$\sqrt{x^2} = \pm\sqrt{48}$$
$$x = \pm\sqrt{48}$$

Put the radical in simplest form.

$$x = \pm\sqrt{16 \cdot 3}$$
$$x = \pm\sqrt{16}\sqrt{3}$$
$$x = \pm 4\sqrt{3}$$

That's a nice method for solving equations that can be written as $x^2 = $ a constant, but if that were all it could do, it wouldn't be much help. The square root method is not the most powerful method and it won't solve every quadratic equation, but it can deal with some more complicated equations. It can be used to solve any equation that is in the form $\left(\text{some expression}\right)^2 = $ a constant.

If you need to solve the equation $(x+3)^2 = 81$, you can use the square root method with a little bit of extra work. Take the square root of both sides.

$$(x+3)^2 = 81$$
$$\sqrt{(x+3)^2} = \pm\sqrt{81}$$
$$x+3 = \pm 9$$

Just as taking the square root of x^2 will undo the squaring and get us back to x, taking the square root of $(x+3)^2$ will undo the squaring of the quantity. This leaves you not with $x = $ a number, but with $x+3$ equal to 9 or to -9. The extra step will be translating that information into two little equations, each of which will give you one solution. It looks like this.

$$(x+3)^2 = 81$$
$$\sqrt{(x+3)^2} = \pm\sqrt{81}$$
$$x+3 = \pm 9$$

$x+3 = 9$	$x+3 = -9$
$x = 6$	$x = -12$

The solutions of $(x+3)^2 = 81$ are $x = 6$ or $x = -12$. You can check that by substituting each solution, one at a time, into the original equation.

$$(x+3)^2 = 81 \qquad\qquad (x+3)^2 = 81$$
$$(6+3)^2 = 81 \qquad\qquad (-12+3)^2 = 81$$
$$9^2 = 81 \qquad\qquad\quad (-9)^2 = 81$$

 TIP

If the $(\text{expression})^2$ equals a negative number, you can't take the square root of both sides because negative numbers don't have square roots, at least, not in the real numbers. If you find that $(\text{expression})^2 =$ a negative number, the equation has no real solution.

Here's another example of an equation that can be solved by the square root method: $4x^2 + 20x + 25 = 16$. You can use the square root method on any equation in the form $(\text{some expression})^2 =$ a constant, and $4x^2 + 20x + 25 = 16$ certainly doesn't look like that. If you remember the special forms you learned when we talked about multiplying polynomials and about factoring, you might recognize $4x^2 + 20x + 25$ as a perfect square trinomial. You can rewrite $4x^2 + 20x + 25 = 16$ as $(2x+5)^2 = 16$ and that will let you use the square root method.

$$4x^2 + 20x + 25 = 16$$
$$(2x+5)^2 = 16$$
$$\sqrt{(2x+5)^2} = \pm\sqrt{16}$$
$$2x+5 = \pm 4$$

Now break that up into two equations.

$$2x+5 = \pm 4$$

$$2x+5 = 4 \qquad\qquad 2x+5 = -4$$
$$2x = -1 \qquad\qquad 2x = -9$$
$$x = -\frac{1}{2} \qquad\qquad x = -\frac{9}{2}$$

 ALGEBRA TRAP

Because your first exposure to quadratic equations is usually equations of the form $x^2 =$ a constant, where the solutions are the positive and negative of the same number, it's easy to think that all quadratics have solutions that are the positive and negative of the same number. That's not the case. You should expect two solutions, but don't solve for one and assume the other is its opposite. Take your time and find both solutions.

The two solutions of $4x^2 + 20x + 25 = 16$ are $x = -\frac{1}{2}$ and $x = -\frac{9}{2}$.

CHECK POINT

Solve each equation by the square root method.

1. $x^2 = 81$

2. $2x^2 = 50$

3. $x^2 - 10 = 6$

4. $(x-7)^2 = 4$

5. $(x+3)^2 - 1 = 8$

6. $x^2 - 6x + 9 = 1$

7. $x^2 + 121 = 22x$

8. $(3x-4)^2 = 49$

9. $4x^2 + 4x + 1 = 36$

10. $(5x-8)^2 = 18$

The Quadratic Formula

Many years ago, people realized that although there were other methods for solving quadratic equations, they were a lot of work, and often very messy arithmetic.

Luckily for us, some patient soul got an idea. What would happen if you took the general pattern of a quadratic equation, $ax^2 + bx + c = 0$, right from the definition, and went through all the steps of the other method, with the a, b, and c still in there? We won't go through all the steps, because, as you can imagine, it is messy. But that patient soul did go through all the steps, and here's how it turned out.

$$x = \frac{-b \pm \sqrt{b^2 - 4ac}}{2a}$$

That equation, the solution that patient person arrived at has come to be known as the *quadratic formula*. That formula is our good fortune. It's an easy way to solve a quadratic equation, without worrying about whether it's a perfect square, or whether it can be made a perfect square. We just need to match our equation up with $ax^2 + bx + c = 0$, identify a, b, and c, and plug those values into the quadratic formula.

 **DEFINITION**

The **quadratic formula** is $x = \frac{-b \pm \sqrt{b^2 - 4ac}}{2a}$. It can be used to find solutions for any quadratic equation of the form $ax^2 + bx + c = 0$ by substituting the values of a, b, and c and simplifying.

To solve the equation $x^2 - 3x + 7 = 5$, begin by gathering all the terms on the left side, equal to 0. Do that by subtracting 5 from both sides.

$$x^2 - 3x + 7 = 5$$
$$x^2 - 3x + 2 = 0$$

Compare $x^2 - 3x + 2 = 0$ to $ax^2 + bx + c = 0$ to identify a, b, and c.

$$ax^2 + bx + c = 0$$
$$\downarrow \quad \downarrow \quad \downarrow$$
$$x^2 - 3x + 2 = 0$$
$$a = 1 \quad b = -3 \quad c = 2$$

Now plug the values into the quadratic formula and replace the a with 1, the b with -3, and the c with 2.

$$x = \frac{-b \pm \sqrt{b^2 - 4ac}}{2a}$$

$$x = \frac{-(-3) \pm \sqrt{(-3)^2 - 4(1)(2)}}{2(1)}$$

Simplify as much as possible.

$$x = \frac{-(-3) \pm \sqrt{(-3)^2 - 4(1)(2)}}{2(1)}$$

$$x = \frac{3 \pm \sqrt{9 - 8}}{2}$$

$$x = \frac{3 \pm \sqrt{1}}{2}$$

$$x = \frac{3 \pm 1}{2}$$

Break into two equations.

$$x = \frac{3 \pm 1}{2}$$

$$x = \frac{3 + 1}{2} \qquad x = \frac{3 - 1}{2}$$

$$x = \frac{4}{2} \qquad x = \frac{2}{2}$$

$$x = 2 \qquad x = 1$$

While that may not seem all that much simpler, the quadratic formula does have the advantage of solving any quadratic, and it's really just plug in and simplify.

 TIP

If your quadratic equation does not have an x-term, a first power term, then $b = 0$ in the quadratic formula. If there is no constant term, just an x^2 term and an x-term, then $c = 0$.

Here's another example. Solve $2x^2 - 5x - 2 = 0$.

$$2x^2 - 5x - 2 = 0$$
$$a = 2 \quad b = -5 \quad c = -2$$
$$x = \frac{-b \pm \sqrt{b^2 - 4ac}}{2a}$$
$$x = \frac{-(-5) \pm \sqrt{(-5)^2 - 4(2)(-2)}}{2(2)}$$
$$x = \frac{5 \pm \sqrt{25 - (-16)}}{4}$$
$$x = \frac{5 \pm \sqrt{41}}{4}$$

When the solutions are irrational, as they are here, you just need to be sure the radical is in simplest form, and you're done. The two solutions are $x = \dfrac{5 + \sqrt{41}}{4}$ and $x = \dfrac{5 - \sqrt{41}}{4}$.

 CHECK POINT

Solve each quadratic equation by using the quadratic formula.

11. $x^2 + 4x - 21 = 0$

12. $t^2 = 10 - 3t$

13. $y^2 - 4y = 32$

14. $x^2 = 6 + x$

15. $9 = 6w + w^2$

16. $x^2 + 7x + 15 = x$

17. $4x^2 - 2 = x + 1$

18. $3x^2 - 1 = 2x$

19. $x^2 + 6x = 4$

20. $x^2 + 4x - 2 = 0$

Using the Discriminant

The portion of the quadratic formula under the radical, $b^2 - 4ac$, is called the *discriminant*. The name comes from the definition of discriminate: recognize a distinction. The discriminant allows us to distinguish between quadratic equations that have, as expected, two distinct real solutions, quadratic equations that have no solution, and quadratic equations that appear to have only one solution.

I say "appear to" because they actually have two solutions that are the same. This is sometimes called a *double root*. The equation $x^2 = 0$ is a simple example of an equation with a double root. Technically, it has two solutions: $x = 0$ and $x = 0$. That really only looks like one solution.

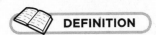 **DEFINITION**

> The **discriminant** is the portion of the quadratic formula that appears under the radical, $b^2 - 4ac$. It can be used to tell how many solutions (and what type of solutions) a quadratic equation will produce.
>
> A **double root**, or double solution, is a solution that occurs twice for the same equation, because the equation was a perfect square trinomial.

If the discriminant is a positive number, the equation will have two distinct real solutions. If the discriminant is positive and a perfect square, those solutions will be rational numbers. If it's positive but not a perfect square, the solutions will be irrational.

If the discriminant is zero, the equation will have a double root, and therefore appear to have only one solution.

If the discriminant is a negative number, it will be impossible to take its square root, so the equation will have no solution.

Discriminant $b^2 - 4ac$	Solutions	Example
Positive and perfect square	Two distinct real, rational solutions	$x^2 - 5x + 4 = 0$ has $b^2 - 4ac = 25 - 4(1)(4) = 9$. Solutions $x = 1$ and $x = 4$
Positive, not a perfect square	Two distinct real, irrational solutions	$x^2 - 5x + 3 = 0$ has $b^2 - 4ac = 25 - 4(1)(3) = 13$. Solutions $x = \dfrac{5 \pm \sqrt{13}}{2}$
Zero	One solution (a double root)	$x^2 - 4x + 4 = 0$ has $b^2 - 4ac = 16 - 4(1)(4) = 0$. Solutions $x = 2$ and $x = 2$
Negative	No real solution	$x^2 - 2x + 4 = 0$ has $b^2 - 4ac = 4 - 4(1)(4) = -12$. There is no real solution.

CHECK POINT

Find the discriminant.

21. $x^2 - 8x + 12 = 0$

22. $x^2 - 3x - 9 = 0$

23. $2x^2 + 5x + 4 = 0$

24. $x^2 + 8x + 16 = 0$

25. $x^2 - 9 = 0$

Use the discriminant to determine the number of solutions, but do not solve.

26. $x^2 - 5x + 4 = 0$

27. $2x^2 - 5x - 4 = 0$

28. $9x^2 + 4 = 0$

29. $x^2 - 8x + 16 = 0$

30. $3x^2 - 2x - 1 = 0$

In the previous sections of this chapter, we looked at various methods for solving quadratic equations. The square root method is quick and easy in the right situations, and the quadratic formula can be counted on to solve any quadratic equation. Even the quadratic formula can lead to some messy arithmetic at times, however, and that's why many people find our last method, solving by factoring, to be their favorite.

Like the other methods, factoring is not perfect. Not every quadratic expression can be factored, as you've seen. Some equations that can be solved by the quadratic formula just won't yield to factoring, but when factoring does work, it's often the fastest, simplest method.

The Factoring Method

The factoring method is based on something called the zero product principle. That's a very formal name for something you likely already know. Think of two numbers that, when multiplied, produce a product of 0. Got them? Zero and what else? Any time two numbers multiply to 0, at least one of them must have been 0. That's the zero product principle. If $ab = 0$, then either $a = 0$ or $b = 0$ or both.

We're going to pull together that principle and what we know about factoring to solve many quadratic equations. Suppose you want to solve $x^2 + 11x + 28 = 0$. You could use the quadratic formula, but the numbers are going to get large. So let's try this. If you can write $x^2 + 11x + 28$ as a product and factor it, you'll have a product that equals zero, and that means one or both of the factors are 0.

There is a 0 on the right side of the equation. That's important. Factoring won't do any good if you don't have that. Let's factor.

$$x^2 + 11x + 28 = 0$$
$$(x+4)(x+7) = 0$$

According to the zero product property, if $(x+4)$ and $(x+7)$ multiply to 0, then either $x+4=0$ or $x+7=0$ or both. Let's look at those two possibilities.

$$x+4=0 \qquad x+7=0$$
$$x=-4 \qquad x=-7$$

The two solutions of $x^2 + 11x + 28 = 0$ are $x=-4$ or $x=-7$.

When the quadratic expression is factorable, this is often the quickest way to solve the equation. Any type of factoring—greatest common factor, FOIL factoring, special forms—can be used, but the quadratic expression must be equal to zero before you begin to factor. Here's another example.

Solve $5x^2 = 40x$. Start by getting all the non-zero terms on the left side, equal to 0.

$$5x^2 = 40x$$
$$5x^2 - 40x = 0$$

Now we can factor out a common factor of $5x$ from both terms.

$$5x^2 - 40x = 0$$
$$5x(x-8) = 0$$

Use the zero product principle to make two equations by setting each factor equal to 0.

$$5x=0 \qquad x-8=0$$
$$x=0 \qquad x=8$$

The solutions of $5x^2 = 40x$ are $x=0$ or $x=8$.

Let's take a look at an example that doesn't look like a quadratic at first. Solve the radical equation $\sqrt{2x+13} = x+5$. Remember that our strategy for solving radical equations is to isolate the radical and square both sides. The radical is isolated, but squaring $x+5$ is going to result in a squared term and leave us with a quadratic equation.

$$\sqrt{2x+13} = x+5$$
$$\left(\sqrt{2x+13}\right)^2 = (x+5)^2$$
$$2x+13 = x^2 + 10x + 25$$

Gather all the non-zero terms on one side, equal to 0, and factor.

$$0 = x^2 + 8x + 12$$
$$0 = (x+2)(x+6)$$

Set each factor equal to 0 and solve.

$$x + 2 = 0 \qquad x + 6 = 0$$
$$x = -2 \qquad x = -6$$

Don't forget that when you solve radical equations, even radical equations that turn into quadratic equations, you must check your solutions. Extraneous solutions are always possible.

$$\sqrt{2x+13} = x+5 \qquad\qquad \sqrt{2x+13} = x+5$$
$$\sqrt{2(-2)+13} = -2+5 \qquad \sqrt{2(-6)+13} = -6+5$$
$$\sqrt{-4+13} = 3 \qquad\qquad \sqrt{-12+13} = -1$$
$$\sqrt{9} = 3 \qquad\qquad\qquad \sqrt{1} = -1$$
$$3 = 3 \qquad\qquad\qquad 1 \neq -1$$

One of the solutions checks, but the other does not. Reject the extraneous solution $(x = -6)$ and the solution is $x = -2$. The quadratic equation that was created when you squared both sides does have two solutions, but only one of them is a solution of the original radical equation.

CHECK POINT

Solve each quadratic equation by factoring, if possible. If the expression is not factorable, use the quadratic formula.

31. $x^2 + 5x + 6 = 0$

32. $x^2 + 12 = 7x$

33. $x^2 - 3x - 4 = 0$

34. $x^2 - 6x + 5 = 0$

35. $x + 2 = 3x^2$

36. $6x^2 - 5x - 6 = 0$

37. $x^2 + 3x = 8 + x$

38. $x^2 + x - 20 = 0$

39. $x^2 + 3x = 0$

40. $3x^2 - 3x + 1 = x^2$

The Least You Need to Know

- Quadratic equations in the form $(\text{some expression})^2 = \text{a constant}$, and those that can be conveniently put in that form, can be solved by taking the square root of both sides.

- Any quadratic equation $ax^2 + bx + c = 0$ can be solved by the quadratic formula $x = \dfrac{-b \pm \sqrt{b^2 - 4ac}}{2a}$.

- The part of the quadratic equation known as the discriminant $b^2 - 4ac$ tells how many solutions to expect. If it's positive, there will be two solutions. If it's zero, there will be one solution, and if the discriminant is negative, the equation has no real solution.

- If a quadratic expression is factorable and equal to zero, the equation can be solved by factoring and setting each factor equal to zero.

Quadratic Functions

Have you ever seen something about to fall and wondered if you have enough time to catch it before it hits the ground? Have you ever tossed a ball in the air and caught it as it came down? Quadratic functions can be used to describe those situations.

The motion of objects dropped or thrown is just one situation that can be described by quadratic functions. In the previous chapter, we learned a variety of ways to solve equations in the form $ax^2 + bx + c = 0$. In this chapter, we'll look at functions of the form $f(x) = ax^2 + bx + c$ or $y = ax^2 + bx + c$. We'll investigate quick ways to graph them, and how to find the equation that matches a graph. Once we have those skills, we can return to vertical motion problems and gather some information about those dropped and thrown objects.

In This Chapter

- Graphing quadratic functions by building a table of values
- Using clues in the equation to graph parabolas quickly
- Graphing quadratic inequalities
- Finding the equation of a parabola from its graph
- Using quadratic functions to solve vertical motion problems

Graphing Quadratic Functions

A picture, it is said, is worth a thousand words, and when it comes to functions, that may well be true. You can quickly learn a lot by looking at the graph of a quadratic function. Let's start by looking at the simplest possible quadratic function, $y = x^2$, and make a table of values. The result is a picture of the characteristic shape of quadratic graphs. The graph of a quadratic function will always have a cup shape known as a *parabola*.

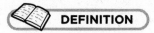 **DEFINITION**

A **parabola** is a cup-shaped graph characteristic of quadratic functions. Technically, it is defined as a set of points each of which is equidistant from a predetermined point and line. This gives a very specific shape to the cup.

Here's the table of values for $y = x^2$.

x	-3	-2	-1	0	1	2	3
y	9	4	1	0	1	4	9
Point	(-3, 9)	(-2, 4)	(-1, 1)	(0, 0)	(1, 1)	(2, 4)	(3, 9)

And when you plot those points and connect them, here's the graph of $y = x^2$.

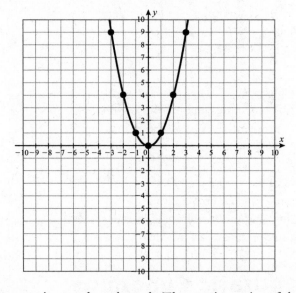

The parabola is a symmetric cup-shaped graph. The turning point of the parabola is called the *vertex*. In this example, the vertex is the origin, (0, 0). A vertical line passing through that vertex is the *axis of symmetry*. Every point on the parabola has a reflection on the parabola. Imagine that the axis of symmetry is a mirror, and a point on the parabola like (2, 4) has a mirror image, (-2, 4). If you were to fold the graph along the *y*-axis, the two sides of the parabola would match each other perfectly.

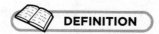 **DEFINITION**

The **vertex** of a parabola is the turning point of the graph. The **axis of symmetry** is an imaginary line passing through the vertex. The two sides of the parabola are reflections of each other across the axis of symmetry.

Here's a table of values and graph for another quadratic function, the function $y = x^2 + 8x + 5$.

x	-3	-2	-1	0	1	2	3
y	-10	-7	-2	5	14	25	38
Point	(-3, -10)	(-2, -7)	(-1, -2)	(0, 5)	(1, 14)	(2, 25)	(3, 38)

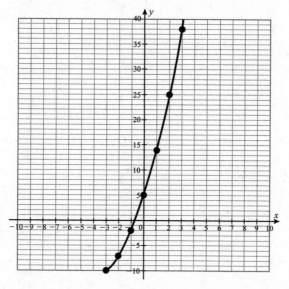

This graph doesn't appear to have that parabola shape. It looks like a line with a little bit of a bend in it. Why isn't it a parabola?

The answer is simple. It is a parabola, but the cup shape has moved and is "off-screen" at the moment. Here's a different view that might make things clearer. This one shows the parabola $y = x^2$ as well as $y = x^2 + 8x + 5$.

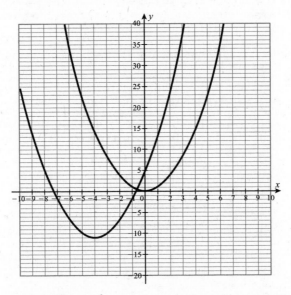

The values for x in the table for $y = x^2 + 8x + 5$ were all on the same side of the parabola, so you can't see the graph turn and form the cup-shape. That could be a problem, so we need to be smarter about making our table.

Vertex and Table of Values

To avoid graphing parabolas that don't look like parabolas, the table of values must include points on both sides of the vertex. To do that, you need to know, before you start graphing, where the vertex is.

The parabola defined by the equation $y = ax^2 + bx + c$ has an axis of symmetry that is a vertical line passing right through the vertex. The equation of that vertical line is $x = -\dfrac{b}{2a}$. For the function we graphed earlier, $y = x^2 + 8x + 5$, $a = 1$ and $b = 8$, so the axis of symmetry is the vertical line $x = -\dfrac{8}{2 \cdot 1}$ or $x = -4$. The vertex is on that vertical line, so the turning point has an x-coordinate of -4.

TIP

If the quadratic equation does not have an x-term, the axis of symmetry is the line $x = 0$, the y-axis, because the value of b in $x = \dfrac{-b}{2a}$ is zero.

To be sure that you see the cup-shape you expect for a parabola, you need to choose some x-values less than -4 and some greater. You could choose -5, -6, and -7 for the points less than -4 and -3, -2, and -1 for the greater points. If you build a table with those values, you'll see the graph turn.

x	-7	-6	-5	-4	-3	-2	-1
y	-2	-7	-10	-11	-10	-7	-2
Point	(-7, -2)	(-6, -7)	(-5, -10)	(-4, -11)	(-3, -10)	(-2, -7)	(-1, -2)

If you had started out with these values, your first look at the graph of $y = x^2 + 8x + 5$ would have looked like this.

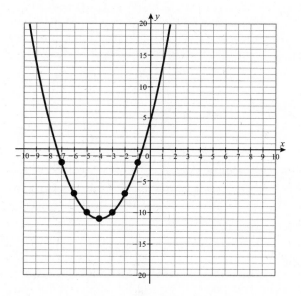

 TIP

The parabola is not made up of lines. It's a curve. After you plot points, don't try to connect the dots with straight lines. Sketch the curve through the points smoothly. It may help to sketch the curve at the vertex first, then extend up (or down) on both sides.

It may seem odd to choose only negative values for x. When we graphed lines, we usually chose both positive and negative values. The nature of the graph of a quadratic function means that it's more important to create a graph that shows the interesting parts of the graph, especially the vertex.

To graph a quadratic function of the form $y = ax^2 + bx + c$,

1. Find the equation of the axis of symmetry using $x = -\dfrac{b}{2a}$. This tells you the x-coordinate of the vertex.

2. Choose a few x-values less than the x-coordinate of the vertex, and a few greater than that coordinate.

3. Build a table of values, with the values you've chosen, including the vertex.

4. Plot the points and connect them with a smooth curve.

 TIP

One principal characteristic of the parabola is that it is symmetric. Every point, except the vertex, has a mirror image on the other side of the axis of symmetry. When you plot your points, expect to see this symmetry. If you don't, it may be time to check your table of values for errors in arithmetic.

Vertex and Intercepts

Solving a quadratic equation can also help you get an appropriate graph of a quadratic function. If you want to graph quadratic functions of the form $y = ax^2 + bx + c$, you can replace y with 0, giving you the quadratic equation $0 = ax^2 + bx + c$, which you know how to solve. Solving it will result in two values for x, each of which pairs up with the 0 you put in for y, to give you two points on the graph.

 TIP

Finding the x-intercepts by solving the quadratic equation is helpful for getting a graph quickly. If the quadratic equation has only one solution, however, it will only have one x-intercept because its vertex just touches the x-axis. If it has no solution, it will float above or below the x-axis and never cross it, so there will be no x-intercepts. In either of those situations, you'll need to create a table of values.

Specifically, those two points, because their y-coordinate is 0, are the x-intercepts of the graph. You'll have (in most cases) two points on the x-axis that you know the graph will pass through. If you also find the vertex, and find the y-intercept by substituting 0 for x, you may be able to graph the function without a table of values. Let's look at an example.

To graph $y = x^2 + 10x + 21$, you need to extract three pieces of information: the vertex, the x-intercepts, and the y-intercept.

For the x-coordinate of the vertex, use $x = -\dfrac{b}{2a}$ with $a = 1$ and $b = 10$. The x-coordinate of the vertex is $x = -\dfrac{10}{2 \cdot 1} = -5$. Find the y-coordinate of the vertex by plugging -5 into the equation, just as you would do if you were making a table of values. $y = x^2 + 10x + 21 = (-5)^2 + 10(-5) + 21 = 25 - 50 + 21 = -4$. The vertex is (-5, -4).

To find the x-intercepts of the graph, solve the equation $0 = x^2 + 10x + 21$ by factoring.

$$x^2 + 10x + 21 = 0$$
$$(x + 3)(x + 7) = 0$$

$$x + 3 = 0 \qquad\qquad x + 7 = 0$$
$$x = -3 \qquad\qquad\quad x = -7$$

The x-intercepts are (-3, 0) and (-7, 0).

Finally, to find the y-intercept, replace x with 0 in $y = x^2 + 10x + 21$. $y = 0^2 + 10 \cdot 0 + 21 = 21$. The y-intercept is (0, 21).

 TIP

The y-intercept of the graph of a quadratic equation is always the constant term.

Let's get those points on a graph.

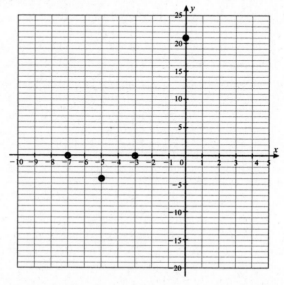

Can you see the cup-shape of the parabola forming? You can use the fact that parabolas have an axis of symmetry to predict that if (0, 21) is on the graph, so is (-10, 21), and that will help make the parabola clearer. Connect the points with a smooth curve and you're done.

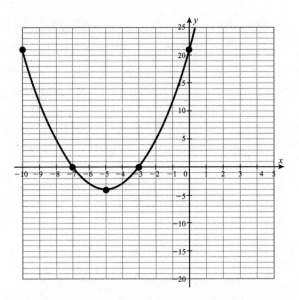

Let's tackle one more example. This time let's graph $y = 4 - x^2$. You're used to working with functions in the form $y = ax^2 + bx + c$, so let's examine this one and make sure you have the correct values for a, b, and c. You can rewrite $y = 4 - x^2$ as $y = -1x^2 + 0x + 4$, and then it's clear that $a = -1$, $b = 0$, and $c = 4$.

When you try to find the vertex, $b = 0$ so $x = -\dfrac{b}{2a} = 0$. The x-coordinate of the vertex is 0 and if you plug that into the equation $y = 4 - x^2$, you find that the y-coordinate of the vertex is 4. The vertex is (0, -4), and coincidentally, (0, -4) is also the y-intercept.

Solve $4 - x^2 = 0$ to find the x-intercepts. You can use the square root method or the factoring method (or the quadratic formula, but that's working too hard). Let's use the square root method and say $4 - x^2 = 0$ so $4 = x^2$ and $x = \pm 2$. The x-intercepts are (2, 0) and (-2, 0).

When you plot those points, you can see the parabola forming, but this time, it's upside down.

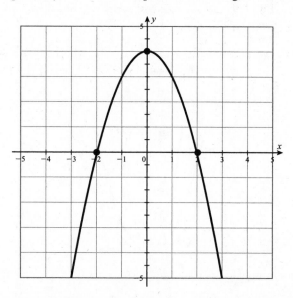

 CHECK POINT

Graph each quadratic function by finding the vertex, the x-intercepts, the y-intercept, and if necessary, making a table of values.

1. $y = x^2 + 4x - 5$

2. $y = x^2 - 2x - 3$

3. $y = 4 - 3x - x^2$

4. $y = x^2 - 8x$

5. $y = 32 + 4x - x^2$

6. $y = x^2 + 4x - 21$

7. $y = 2x^2 - 18$

8. $y = 2x^2 + 10x + 12$

9. $y = -2x^2 - 12x - 18$

10. $y = 4x^2 + 4x + 1$

Graphing Shortcuts

You've already learned two particularly quick methods for graphing linear equations: x- and y-intercepts, and slope and y-intercept. To graph quadratic functions, you need a little more information, because the shape of the graph is more complex. You can use the vertex and the intercepts, but the equation has a few other clues that will help you sketch the graph quickly.

The value of a in $y = ax^2 + bx + c$ gives you two pieces of information. Its sign tells you whether the parabola opens up, forming a cup that would hold liquid, or opens down, spilling the liquid.

If a is positive, the parabola opens up. If a is negative, the parabola opens down. The graph of $y = x^2 + 10x + 21$ opened up because the value of a was +1, but the graph of $y = 4 - x^2$ opened down because of the -1 in the a position.

The value of a also tells about the width of the parabola, whether it's narrow or wide. The larger the absolute value of a, the narrower the parabola will be. Look at the following graph, which shows $y = x^2$, $y = 3x^2$, and $y = \frac{1}{2}x^2$.

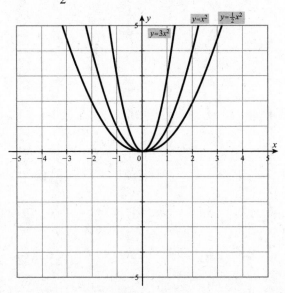

It's nice to know whether you're going to have a narrow or a wide parabola, but that's not much help unless you know how wide or how narrow. In fact, the value of a can help you answer that, if you know an old trick called the odd number rule.

Think about the odd numbers: 1, 3, 5, and so on. Then look at the graph of $y = x^2$, and imagine putting one finger of each hand on the vertex. Now move one finger one box right and the other one box left. Your fingers are on the points (1, 0) and (-1, 0). How far do you need to move your fingers upward to get to the parabola? One unit up on each side should do it, so move to (1, 1) and (-1, 1). Now move out one unit on each side again, to (2, 1) and (-2, 1). How far to the parabola now? Three units up? Can you guess what will happen if you do it again? You'll have to go 5 units up, and the next time 7, then 9, and on through the odd numbers.

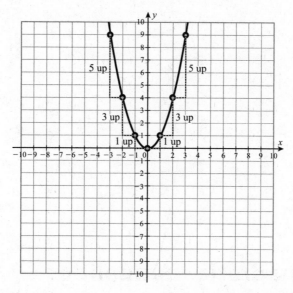

Any time $a = 1$, you'll see this 1, 3, 5, ... pattern. If a is not 1, take the odd numbers and multiply each one by a. So when you graph $y = 3x^2$, you find the vertex, (0, 0), and multiply the odd numbers by 3. $3 \cdot 1$, $3 \cdot 3$, $3 \cdot 5$,... becomes 3, 9, 15, ... You can start from the vertex and count out 1, up 3, out 1, up 9, out 1, up 15, to set points of the graph.

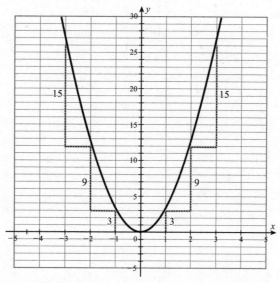

You already know how to find the x-coordinate of the vertex using $x = -\dfrac{b}{2a}$. Finally, c, the constant term, is the y-intercept. If you use the vertex and these helpers, you can often sketch the graph of a parabola without a lot of prep work.

CHECK POINT

Graph each quadratic function by using shortcuts wherever you can. If necessary, make a table of values.

11. $y = 2x^2$

12. $y = 3x^2 - 4$

13. $y = 5 - x^2$

14. $y = x^2 + 6x$

15. $y = x^2 - 2x - 3$

16. $y = 3 - 2x - x^2$

17. $y = 9 - 2x^2$

18. $y = x^2 - 1$

19. $y = 9 - x^2$

20. $y = 2x^2 - 6x$

Graphing Quadratic Inequalities

When you learned to graph linear functions, you also explored linear inequalities, relations that are not functions and are best communicated by showing a graph. Let's use the knowledge, along with what you've just learned, to graph quadratic inequalities, and use those graphs to understand the solution set of the inequality.

The quadratic function $y = 6 + 4x - 2x^2$ can be graphed fairly quickly if you recognize that $a = -2$, $b = 4$, and $c = 6$, and use those to find the vertex and y-intercept. The x-coordinate of the vertex is $x = -\dfrac{4}{2(-2)} = 1$, and the y-coordinate is $y = 6 + 4 \cdot 1 - 2 \cdot 1^2 = 8$. The vertex is (1, 8) and the y-intercept is (0, 6). Then you can make a table or solve $0 = 6 + 4x - 2x^2$ to find the x-intercepts.

$$0 = 6 + 4x - 2x^2$$
$$-2x^2 + 4x + 6 = 0$$
$$-2\left(x^2 - 2x - 3\right) = 0$$
$$-2\left(x + 1\right)\left(x - 3\right) = 0$$
$$x + 1 = 0 \qquad x - 3 = 0$$
$$x = -1 \qquad\quad x = 3$$

Here's the graph of the quadratic function $y = 6 + 4x - 2x^2$.

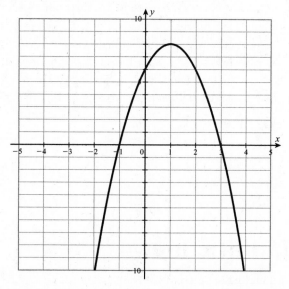

That graph shows the points for which the y-coordinate is equal to $6 + 4x - 2x^2$, but you won't always want to know where y equals the expression. Sometimes you might be interested in the points for which y is less than the expression or greater than the expression. So if the preceding graph is the graph of $y = 6 + 4x - 2x^2$, what would the graph of $y < 6 + 4x - 2x^2$ look like?

Use what you know about linear inequalities to answer the question. Remember that you used a solid line to graph inequalities that included "or equal to" but a dotted line to show strictly greater or strictly less. So the first change you would make to the preceding graph would be to draw the parabola with a dotted line.

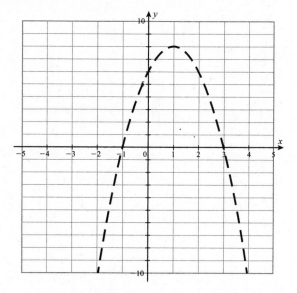

That communicates that although the parabola is a boundary for the points you want, you don't want to actually include the boundary. Where then are the points you want? Which side of the boundary has points with a y-coordinate less than $6+4x-2x^2$? The basic rule you used with lines still holds. For y less than the expression, shade down. For y greater than the expression, shade up. Of course, you can always test a point to see if it is in a region that should be shaded. You want $y<6+4x-2x^2$, so shade down. Check by testing the point (0, 0) in the inequality. $0<6+4\cdot0-2\cdot0^2$, or $0<6$, is a true statement, so you're shading the correct area.

Here's what the final graph of $y<6+4x-2x^2$ looks like.

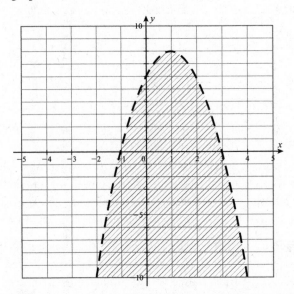

 CHECK POINT

Graph each quadratic inequality.

21. $y<-2x^2$

22. $y>x^2-6$

23. $y\le x^2+4x$

24. $y\ge x^2-4$

25. $y\le x^2-2x-3$

26. $y>x^2+4x-5$

27. $y<4-3x-x^2$

28. $y\ge x^2+6x$

29. $y<x^2+4x-21$

30. $y>2x^2+10x+12$

Finding the Equation of a Parabola

Do you remember how to find the equation of a line? You need to know two points on the line, or one point and the slope. Then you can use the point-slope form of the line, $y - y_1 = m(x - x_1)$, and replace m with the slope and x_1 and y_1 with the coordinates of a point.

That won't work for finding the equation of a parabola, for the obvious reason that a parabola is not a line, and that's the point-slope form of a line. Because the parabola is a more complex shape than a line, just knowing any two points on the graph won't be enough. And a parabola doesn't have a slope. A slope is the rate at which a line rises or falls, each time it moves one unit right. The slope of a line is the same no matter where on the line you measure it, but that's certainly not true of a parabola. Parabolas fall on one side and rise on the other, and they rise or fall more quickly as you move away from the vertex. So what can you do about finding the equation of a parabola from the graph?

As you've graphed parabolas, you've probably noticed that the vertex locates the graph for you, and then the value of a tells you how to shape it. Those two pieces of information—the vertex and the value of a—define the parabola. Luckily, there's a form of the equation of a parabola called the vertex form that allows you to write the equation if you know those two pieces of information.

First, you need to know the vertex. Denote the vertex as (h, k). Instead of point-slope form, $y - y_1 = m(x - x_1)$, which uses (x_1, y_1) as some point on the line, you're going to use (h, k) to represent one very specific point on the parabola. Even if you made that change, writing $y - k = m(x - h)$, you'd still be writing the equation of a line. To create a parabola, you have to have a variable squared, so you need something like this: $y - k = m(x - h)^2$. But m stands for the slope, and a parabola doesn't have a slope, so change that to a, and the vertex form of the equation of a parabola will be $y - k = a(x - h)^2$.

If you're asked to find a parabola with vertex $(4, -7)$ and $a = 3$, you can take the vertex form and plug in your specific numbers.

$$y - k = a(x - h)^2$$
$$y - (-7) = 3(x - 4)^2$$

All that's left to do is simplify.

$$y - k = a(x - h)^2$$
$$y - (-7) = 3(x - 4)^2$$
$$y + 7 = 3(x^2 - 8x + 16)$$
$$y + 7 = 3x^2 - 24x + 48$$
$$y = 3x^2 - 24x + 41$$

That's not usually a difficult job, but there is still one problem. You're not usually just handed the value of a to use, but you need it to write the equation. You may be able to deduce the value of a from the graph, using the odd number trick we learned to help us graph, but that's not something you can depend on. Instead, look for another point on the graph, other than the vertex. For example, if we ask you to find the equation of a parabola that has its vertex at (3, 5) and also passes through the point (7, -27), here's what you'll do. Start as you did before with the vertex form, and plug in the coordinates of the vertex.

$$y - k = a(x - h)^2$$
$$y - 5 = a(x - 3)^2$$

You don't know a, but you do know that the point (7, -27) is on this graph, so $x = 7$ and $y = -27$ must fit this equation. If you plug them in, a will be the only thing you don't know, and you can solve for it.

$$y - k = a(x - h)^2$$
$$y - 5 = a(x - 3)^2$$
$$-27 - 5 = a(7 - 3)^2$$
$$-32 = a(4)^2$$
$$-32 = 16a$$
$$a = -2$$

Now you know the vertex and the value of a. Start again with vertex form, plug in the vertex and a, but not the other point, and simplify.

$$y - k = a(x - h)^2$$
$$y - 5 = -2(x - 3)^2$$
$$y - 5 = -2(x^2 - 6x + 9)$$
$$y - 5 = -2x^2 + 12x - 18$$
$$y = -2x^2 + 12x - 13$$

CHECK POINT

Find the equation of the parabola that fits the given information.

31. Vertex at (4, -3) and $a = 3$

32. Vertex at (-2, 5) and $a = -2$

33. Vertex at (0, 2), passing through (2, -2)

34. Vertex at (-4, 5), passing through (-2, 13)

35. Vertex at (-3, -2), passing through (-2, 2)

Find the equation of the parabola shown.

36.

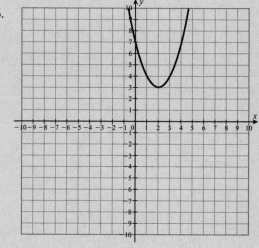

37.

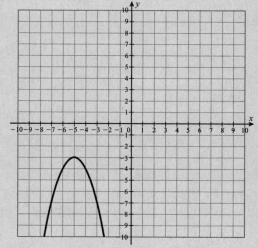

38.

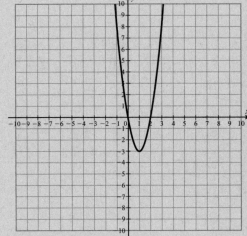

39.

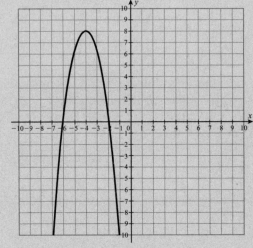

40.

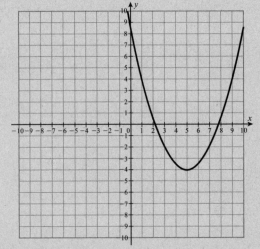

Vertical Motion

One of the most common applications of quadratic functions is as a way to examine the behavior of an object that is dropped or thrown. The function that gives the height of the object at a certain time is a quadratic function with quite a few pieces of information.

$$h(t) = -\frac{1}{2}gt^2 + vt + s$$

$h(t)$ represents the height of the object at some time, t. It's important to be clear on the units of measurement. Feet? Meters? Seconds? Minutes? You can set up an equation using customary units or metric units but you must be consistent and you must know which you're using.

The constant g is the acceleration due to gravity, the earth's pull on the object. This will be in feet per second per second or some similar unit. It's a constant, but the value will depend on whether you're measuring in customary or metric. In customary units, $g \approx 32$ feet per second squared, but in metric units, $g \approx 10$ meters per second squared.

THINK ABOUT IT

In the vertical motion equation, the coefficient of the squared term represents the effect of gravity on the object. It's always negative because gravity pulls the object down, and so reduces its height.

The variable t represents the time since the object was dropped or thrown. It's the moment at which you want to freeze the action and see where the object is.

We're using v to represent the initial velocity of the object, the force that the person throwing the object exerts on it. If the object is dropped, $v = 0$. If the object is thrown upward, v is positive, but if it's thrown downward, v will be a negative number.

Finally, s will be the symbol for the starting height of the object. Was it dropped from the roof of a building? The starting height is the height of the building. If the object is thrown, the starting height is truly the height of your hand when you release it, but often we act as though it was thrown "from the ground," and make $s = 0$. That's actually unlikely to be true—how could you throw it if it were truly on the ground?—but it is close enough in many cases.

Let's look at a couple of examples. If an object is dropped from the top of a 150 foot tower, what is its height after 1 second?

You're working with feet and seconds, so use $g = 32$ feet per second squared, and because the object is dropped, $v = 0$. The starting height is $s = 150$. Plug all the information in, and just this once, we'll leave the units in there so you can see what happens.

$$h(t) = -\frac{1}{2}gt^2 + vt + s$$

$$h(1\text{ sec}) = -\frac{1}{2} \cdot 32\frac{\text{feet}}{\cancel{\text{second}^2}} \cdot \cancel{(1\text{ second})^2} + 0\frac{\text{feet}}{\cancel{\text{second}}} \cdot 1\ \cancel{\text{second}} + 150\text{ feet}$$

$$h(1) = -16\text{ feet} + 150\text{ feet}$$

$$h(1) = 134\text{ feet}$$

Notice how units of time get cancelled out, leaving just units of height. The first term is the work of gravity, and gravity pulled the object down 16 feet in that 1 second. The object wasn't thrown, so the middle term just disappeared. The starting height of 150 feet is reduced by the 16 foot drop due to gravity.

Here's a little more complicated example. If an object is thrown upward from the ground with an initial velocity of 20 meters per second, how much later will it hit the ground?

The initial velocity is 20 meters per second, so use $g = 10$ meters per second squared, and assume that the height of the ground is 0 feet.

$$h(t) = -\frac{1}{2}gt^2 + vt + s$$

$$h(t) = -\frac{1}{2}(10)t^2 + 20t + 0$$

$$h(t) = -5t^2 + 20t$$

You're looking for the time, t, when the height of the object is once again zero.

$$h(t) = -5t^2 + 20t$$

$$0 = -5t^2 + 20t$$

Solve by factoring.

$$-5t^2 + 20t = 0$$

$$-5t(t-4) = 0$$

$$-5t = 0 \qquad t - 4 = 0$$

$$t = 0 \qquad\quad t = 4$$

This tells you that there were two moments when the object was on the ground. When $t = 0$, the object is just being thrown, but the other solution, $t = 4$, tells you that the object will hit the ground again 4 seconds later.

TIP

Vertical motion problems, and any problems that deal with real life situations, are likely to have solutions that are not tidy integer values. Few of them will be solvable by factoring. Be ready to use the quadratic formula.

Type of Question	What to Do
How high is the object at this time?	Plug in g, v, s, and t and simplify.
At what time is the object this high?	Plug in g, v, s, and $h(t)$ and solve for t.
What is the maximum height of the object?	Plug in g, v, and s. Find the vertex of the parabola.

CHECK POINT

For questions 41 through 43: An object is dropped from the top of a 112 foot building.

41. How long will it take to hit the ground?

42. How high is the object one second after it is released?

43. When is the object 76 feet high?

For questions 44 through 47: An object is thrown upward, beginning from a height of 5 feet, with an initial velocity of 160 feet per second.

44. How high is the object after 2 seconds?

45. What is the maximum height the object reaches?

46. When is the object 81 feet off the ground?

47. When does the object hit the ground?

For questions 48 through 50: An object is thrown downward from the top of a 75 meter tower with an initial velocity of 15 meters per second.

48. How high is the object after 0.5 seconds?

49. When is the object 20 meters high?

50. How long does it take to hit the ground?

The Least You Need to Know

- The x-coordinate of the vertex of the parabola with the equation $y = ax^2 + bx + c$ and the equation of its axis of symmetry is $x = \dfrac{-b}{2a}$.

- When graphing quadratic functions, find the vertex, x-intercepts, and y-intercepts to locate the parabola.

- Use the sequence of odd numbers, multiplied by a, to help set the shape of the parabola. Make a table of values if necessary.

- For quadratic inequalities, use dotted lines for strict inequalities and solid line for "or equal to" inequalities. Shade down if y is less than the quadratic expression, and shade up if y is greater.

- For vertical motion problems, use the quadratic function $h(t) = -\dfrac{1}{2}gt^2 + vt + s$ for the height of an object thrown or dropped. In the function, h is the height, t is the time, g is the acceleration due to gravity, v is the initial velocity, and s is the starting height.

- If you're working in customary units, use $g = 32$ feet per second squared. If you're using metric units, $g \approx 10$ meters per second squared.

Rational Expressions

One of the keys to arithmetic is a firm understanding of the place value system that is the structure of whole numbers. One of the keys to algebra is an understanding of polynomials, which have a structure that parallels a place value system. If we write the number 4,829 in expanded form, it's $4 \cdot 10^3 + 8 \cdot 10^2 + 2 \cdot 10 + 9$. If we replace 10 with a variable, we have a polynomial, $4x^3 + 8x^2 + 2x + 9$.

In arithmetic, once we have a sturdy grasp of the whole numbers, we usually move on to fractions. In this chapter and the next, we're going to work with some algebraic fractions. We'll look at their construction, their simplest form, and their arithmetic. Along the way, we'll use much of what we know about polynomials, especially factoring.

Domain of a Rational Expression

The algebraic fractions we're talking about are properly called *rational expressions*. Rational here, like in rational numbers, comes from the word *ratio*. These expressions are ratios of two polynomials, with the warning that the polynomial in the denominator cannot equal zero.

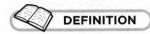

DEFINITION

A **rational expression** is a quotient of two polynomials, provided that the polynomial in the denominator is not zero.

Because the term *polynomial* includes expressions as simple as constants, and others that have many terms, rational expressions can be as simple as $\frac{1}{x}$ or more complicated, like $\frac{x^3 - 4x^2 + 8x + 7}{2x^4 - 9x^3 + 2x^2 - x - 1}$. Technically, all rational numbers are rational expressions and all polynomials are as well, because we can write $\frac{x^2 - 4x + 2}{1}$, but we don't usually bother with that.

ALGEBRA TRAP

You'll sometimes hear rational expressions called algebraic fractions, but the two terms are not identical. There are other algebraic fractions that don't quite fit the definition of rational expressions because their numerator or denominator (or both) are not polynomials. Expressions like $\frac{x^2 - 5x + 7}{\sqrt{x-9}}$ have some properties in common with rational expressions but are often more difficult to work with, because the tactics used with polynomials can't be used here. Don't make the mistake of thinking every algebraic fraction is a rational expression.

The warning that the denominator cannot equal zero is attached to any fraction, because division by zero is impossible. In rational numbers, a zero denominator is easy to spot, but when working with rational expressions, a polynomial denominator can take different values for different values of the variable. Certain values, if substituted for the variable, could make the denominator polynomial equal zero.

It may not always be obvious what values of the variable create a zero denominator. Sometimes it's easy. The denominator of $\frac{3}{x}$ is zero when $x = 0$, but the denominator of $\frac{6}{x-3}$ is not zero when $x = 0$. If you plug $x = 0$ into $\frac{6}{x-3}$, you get $\frac{6}{0-3} = \frac{6}{-3} = -2$, which is perfectly fine. Make a habit of examining each denominator, and asking what values of the variable will make that denominator equal zero.

To answer that question, you need to solve an equation. To determine which value of x will make the denominator of $\frac{6}{x-3}$ equal zero, solve the equation $x - 3 = 0$. That tells you that $x = 3$ is the value you cannot allow. To find the unacceptable values of the variable for the rational expression $\frac{x+4}{x^2 - 3x + 2}$, you need to solve the equation $x^2 - 3x + 2 = 0$. That will take a little bit more work.

$$x^2 - 3x + 2 = 0$$
$$(x-2)(x-1) = 0$$
$$x - 2 = 0 \qquad x - 1 = 0$$
$$x = 2 \qquad x = 1$$

There are two values that will cause the denominator of $\dfrac{x+4}{x^2-3x+2}$ to become zero, $x = 1$ and $x = 2$. Avoid both of those.

The values of the variable that may be substituted for the variable in any expression or function form a set called the *domain* of the expression. For many expressions, like polynomials, the domain is all real numbers. You can replace the variable in a polynomial with any value, and nothing troublesome will happen. With rational expressions, however, there are troublesome values more often than not. The domain of $\dfrac{x+4}{x^2-3x+2}$ is all real numbers except 1 and 2. The simplest way to show this is to tell what's not allowed. If you see $\dfrac{x+4}{x^2-3x+2}, x \neq 1, x \neq 2$, it means that the domain cannot include 1 or 2.

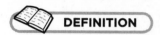 **DEFINITION**

The **domain** of a rational expression is the set of all real numbers that can be substituted for the variable without making the denominator equal to zero.

To find the domain of the rational expression $\dfrac{3x-7}{x^2-9}$, first solve $x^2 - 9 = 0$ to find the values that must be excluded. That will tell you the domain cannot include 3 or -3. The domain of $\dfrac{3x-7}{x^2-9}$ is all real numbers except 3 and -3.

To work with the rational expression $\dfrac{2x-5}{x^2+1}$, first think about its domain, which is all real numbers except anything that will make the denominator zero. When you try to solve $x^2 + 1 = 0$, however, you will find that $x^2 = -1$ and that has no solution in the real numbers. So $\dfrac{2x-5}{x^2+1}$ is one of those unusual rational expressions whose domain is all real numbers.

CHECK POINT

Find the domain of each rational expression.

1. $\dfrac{5}{2x}$

2. $\dfrac{3}{x-7}$

3. $\dfrac{x+1}{x+8}$

4. $\dfrac{2x-5}{3x-9}$

5. $\dfrac{2}{x^2-1}$

6. $\dfrac{4x-1}{x^2+4x+4}$

7. $\dfrac{5x-3}{4x+12}$

8. $\dfrac{2x-5}{x^2+5x}$

9. $\dfrac{x^2-3}{x^2+5x+6}$

10. $\dfrac{7x-1}{x^2+4}$

Simplifying Rational Expressions

Rational expressions are algebraic fractions, and much of what you know about fractions will translate to rational expressions. There will be some changes, some complications, but the essential ideas will be much the same.

One of the first things we learn about fractions is that, although they may actually be two different names for the same number, it's a lot easier to work with $\dfrac{1}{2}$ than with $\dfrac{287}{574}$. Putting a fraction in its simplest form makes both the computation and the communication easier. When working with rational expressions, you'll also want to have the expressions in their simplest form.

Although you may think of simplifying a fraction as a process of dividing the numerator and denominator by the same number, the reasoning behind it is a little more involved. To simplify $\dfrac{84}{168}$, you're actually factoring 84 and factoring 168.

$$\frac{84}{168} = \frac{2 \cdot 2 \cdot 3 \cdot 7}{2 \cdot 2 \cdot 2 \cdot 3 \cdot 7}$$

Then look for factors that appear in both and cancel them out.

$$\frac{84}{168} = \frac{2 \cdot \cancel{2} \cdot \cancel{3} \cdot \cancel{7}}{2 \cdot 2 \cdot \cancel{2} \cdot \cancel{3} \cdot \cancel{7}} = \frac{1}{2}$$

To simplify a rational expression, you'll use a similar process. First factor the numerator and factor the denominator, if they can be factored. Then find and cancel matching factors in the numerator and denominator. Here's an example.

To simplify $\dfrac{x^2 - 4}{x^2 + 3x + 2}$, first factor each polynomial.

$$\frac{x^2 - 4}{x^2 + 3x + 2} = \frac{(x+2)(x-2)}{(x+2)(x+1)}$$

We can see $x + 2$ in both the numerator and denominator, so we'll cancel that out.

$$\frac{x^2 - 4}{x^2 + 3x + 2} = \frac{\cancel{(x+2)}(x-2)}{\cancel{(x+2)}(x+1)} = \frac{x-2}{x+1}, \ x \neq -1, -2$$

ALGEBRA TRAP

When you're simplifying a rational expression, it's tempting to start crossing out anything that looks alike. Remember that "canceling" means dividing the numerator and denominator by the same number or expression. To divide properly, you need to factor the numerator and denominator and cancel factors. $\dfrac{2 \cdot 3}{2 \cdot 5} = \dfrac{3}{5}$ but $\dfrac{2+3}{2+5} \neq \dfrac{3}{5}$. Cancel factors, not terms.

When all the factors that appear in both numerator and denominator are canceled, what remains is the simplest form of the rational expression.

TIP

Determine the domain before you simplify or do any arithmetic. The domain is the set of numbers that replace the variable in the expression you're given, without making the denominator equal zero. $\dfrac{4x}{x(x-3)}$ is undefined for both $x = 0$ and $x = 3$, even though it can be simplified to $\dfrac{4}{x-3}$. The domain is the domain of the unsimplified version, even after simplifying.

The polynomials that make up the rational expression may not always factor, but they may still cancel. The rational expression $\dfrac{5x - 9}{5x^2 + x - 18}$ has a numerator that cannot be factored, but the denominator can be factored. $\dfrac{5x - 9}{5x^2 + x - 18} = \dfrac{5x - 9}{(x+2)(5x-9)}$. One of the factors of the denominator is the polynomial in the numerator. You can rewrite and then cancel to get the simplest form.

$$\frac{5x - 9}{5x^2 + x - 18} = \frac{1\cancel{(5x-9)}}{(x+2)\cancel{(5x-9)}} = \frac{1}{x+2}, \ x \neq -2, \frac{9}{5}$$

CHECK POINT

Simplify each expression and give its domain.

11. $\dfrac{9x^2}{48x^2}$

12. $\dfrac{52x}{13x^2}$

13. $\dfrac{3x+6}{x+2}$

14. $\dfrac{3x+6}{15x-12}$

15. $\dfrac{x^2-6x}{x^2-7x+6}$

16. $\dfrac{x^2-x-42}{x^2-49}$

17. $\dfrac{x^2-8x+15}{x^2-7x+10}$

18. $\dfrac{2x^2+5x+2}{(2x+1)^2}$

19. $\dfrac{x^2-4x}{x^2-16}$

20. $\dfrac{3x^2+6x+3}{6x^2+18x+12}$

Operations with Rational Expressions

Rational expressions, like ordinary fractions, can never have a denominator of zero, and like ordinary fractions, are easiest to deal with when they're in simplest form. Finding the values that must be excluded from the domain to prevent a zero denominator takes a bit of work and factoring the polynomials so that we can put the expression in simplest form will take a few more steps than simplifying a regular fraction. You know how to add, subtract, multiply, and divide fractions. Now it's time to apply and adapt that knowledge so that you can operate with rational expressions.

Multiplication and Division

The basic rule for multiplication of fractions tells you to multiply numerator by numerator and denominator by denominator and then simplify if necessary. That rule will work with rational expressions as well, but you may not want to depend on that to handle every multiplication of rational expressions.

For this problem, it's a workable plan.

$$\frac{5}{x}\cdot\frac{x}{x+1}=\frac{5\cancel{x}}{\cancel{x}(x+1)}=\frac{5}{x+1}$$

For this multiplication, it may feel less comfortable.

$$\frac{x^2+5x+3}{x^2-16}\cdot\frac{x^2+8x+16}{x^3+5x^2+3x}$$

Have you got some scratch paper handy? Multiplying numerator by numerator and denominator by denominator is going to take a lot of work, and when that's done, you'll need to figure out how to factor the results in order to simplify the product.

If you're starting to feel like this is a job you don't want to take on, you're not alone. Just so you can see it, here's what that task would look like, but you don't know how to factor those polynomials, and you aren't expected to, and just about anyone would look at this problem and ask if there's a better way.

$$\frac{x^2+5x+3}{x^2-16} \cdot \frac{x^2+8x+16}{x^3+5x^2+3x} = \frac{x^4+13x^3+59x^2+104x+48}{x^5+5x^4-13x^3-88x^2-48x}$$

$$= \frac{\cancel{(x+4)}(x+4)\cancel{(x^2+5x+3)}}{x\cancel{(x+4)}(x-4)\cancel{(x^2+5x+3)}}$$

$$= \frac{x+4}{x(x-4)}$$

$$= \frac{x+4}{x^2-4x}$$

The good news is that there is a better way, and you already know the heart of it. It's very similar to the way you simplified rational expressions, and it's a technique you've used in working with fractions. You'll simplify before multiplying. You'll factor every polynomial that can be factored, and do as much canceling as possible. Only after that will you multiply.

Let's look at it with a simpler example first, and then we'll come back to the monster we looked at earlier. Let's multiply $\dfrac{x^2+5x+6}{x^2-1} \cdot \dfrac{x^2-2x+1}{x^2-9}$.

First, factor each of the polynomials.

$$\frac{x^2+5x+6}{x^2-1} \cdot \frac{x^2-2x+1}{x^2-9} = \frac{(x+2)(x+3)}{(x+1)(x-1)} \cdot \frac{(x-1)(x-1)}{(x+3)(x-3)}$$

Now you want to cancel a factor from one of the numerators with a factor from one of the denominators. The factors can come from the same fraction or one from each fraction, but it must be one from a numerator and one from a denominator. Cancel the $x+3$ in the first numerator with the $x+3$ in the second denominator, and cancel the $x-1$ in the first denominator with one of the $x-1$ factors in the second numerator.

$$\frac{x^2+5x+6}{x^2-1} \cdot \frac{x^2-2x+1}{x^2-9} = \frac{(x+2)\cancel{(x+3)}}{(x+1)\cancel{(x-1)}} \cdot \frac{\cancel{(x-1)}(x-1)}{\cancel{(x+3)}(x-3)}$$

Now multiply the factors that haven't been cancelled in the numerator, and do the same with the remaining factors in the denominator. If you did all the possible canceling in the last step, no further simplifying will be needed.

$$\frac{x^2+5x+6}{x^2-1} \cdot \frac{x^2-2x+1}{x^2-9} = \frac{(x+2)(x+3)}{(x+1)(x-1)} \cdot \frac{(x-1)(x-1)}{(x+3)(x-3)} = \frac{(x+2)(x-1)}{(x+1)(x-3)} = \frac{x^2+x-2}{x^2-2x-3}$$

Let's try that monster problem we saw earlier, but let's do it by factoring and canceling first.

$$\frac{x^2+5x+3}{x^2-16} \cdot \frac{x^2+8x+16}{x^3+5x^2+3x}$$

First, factor. Notice that x^2+5x+3 is not factorable, but that's okay.

$$\frac{x^2+5x+3}{(x+4)(x-4)} \cdot \frac{(x+4)(x+4)}{x(x^2+5x+3)}$$

The second step is to cancel.

$$\frac{x^2+5x+3}{(x+4)(x-4)} \cdot \frac{(x+4)(x+4)}{x(x^2+5x+3)}$$

Finally, multiply the remaining factors.

$$\frac{x^2+5x+3}{(x+4)(x-4)} \cdot \frac{(x+4)(x+4)}{x(x^2+5x+3)} = \frac{x+4}{x(x-4)} = \frac{x+4}{x^2-4x}$$

That tames the monster a bit, doesn't it? Factoring and canceling first should be your usual approach to multiplying rational expressions. For some simpler problems, you may choose to multiply first, and simplify later, but factoring first is usually your best move.

You probably remember that the rule for dividing fractions tells us to multiply by the reciprocal of the divisor. $\frac{5}{12} \div \frac{15}{48} = \frac{5}{12} \cdot \frac{48}{15}$ The same will be true for dividing rational expressions. We'll leave the first expression as it is, change to multiplication, and invert the second rational expression.

You can remember the rule for division as "keep, change, change". Keep the first rational expression as it is, change to multiplication, and change the second rational expression to its reciprocal.

To divide $\frac{x^2+6x}{x^2-7x+12} \div \frac{x^2+4x-12}{x^2-16}$, keep $\frac{x^2+6x}{x^2-7x+12}$ just as it is for now, change the operation sign to multiplication, and replace $\frac{x^2+4x-12}{x^2-16}$ with its reciprocal, $\frac{x^2-16}{x^2+4x-12}$.

$$\frac{x^2+6x}{x^2-7x+12} \div \frac{x^2+4x-12}{x^2-16} = \frac{x^2+6x}{x^2-7x+12} \cdot \frac{x^2-16}{x^2+4x-12}$$

 TIP

When you're dividing rational expressions, change to multiplication by the reciprocal before you start to factor. If you don't, it's easy to get involved in the factoring and forget to invert the divisor.

Once you've rewritten the division problem as an equivalent multiplication problem, it is a multiplication problem, and your tactic is to factor everything you can, cancel, and then multiply.

$$\frac{x^2+6x}{x^2-7x+12} \div \frac{x^2+4x-12}{x^2-16} = \frac{x^2+6x}{x^2-7x+12} \cdot \frac{x^2-16}{x^2+4x-12}$$

$$= \frac{x\cancel{(x+6)}}{\cancel{(x-4)}(x-3)} \cdot \frac{(x+4)\cancel{(x-4)}}{\cancel{(x+6)}(x-2)}$$

$$= \frac{x(x+4)}{(x-3)(x-2)}$$

$$= \frac{x^2+4x}{x^2-5x+6}$$

 CHECK POINT

Perform each multiplication or division.

21. $\dfrac{3(x+5)^2}{15} \cdot \dfrac{5}{x+5}$

22. $\dfrac{x^2-16}{2} \cdot \dfrac{8}{x+4}$

23. $\dfrac{x^2+2x+1}{5x-10} \cdot \dfrac{2x-4}{x^2-1}$

24. $\dfrac{x^2+7x+10}{x^2-2x+1} \cdot \dfrac{x^2-6x+5}{x^2-25}$

25. $\dfrac{x^2-5x+6}{x^2-7x+10} \cdot \dfrac{x^2-4x-5}{x^2-2x-3}$

26. $\dfrac{5x^2-20}{x^2-5x+6} \div \dfrac{6x+12}{3x-5}$

27. $\dfrac{x^2-4}{2x+5} \div \dfrac{2x-4}{4x^2-25}$

28. $\dfrac{x^2+x-42}{x^2-36} \div \dfrac{x^2-x-42}{x^2-49}$

29. $\dfrac{x^2-1}{3x-6} \div \dfrac{6x+6}{8x-16}$

30. $\dfrac{2x^2+5x+2}{6x+3} \div \dfrac{2x^2+x-6}{2x-3}$

Addition and Subtraction with Like Denominators

Denominators are names, labels that tell what sort of thing we have. Numerators tell how many of that thing we have. If we add, or subtract, groups of the same kind of thing, the number changes but the kind of thing doesn't.

When fractions have the same denominator, like $\frac{5}{7}$ and $\frac{1}{7}$, adding or subtracting only requires adding or subtracting the numerators and keeping the denominator as it is. $\frac{5}{7}+\frac{1}{7}=\frac{6}{7}$ and $\frac{5}{7}-\frac{1}{7}=\frac{4}{7}$. The same is true of rational expressions. If the denominators are the same, you only need to add or subtract the numerators, and keep the denominators the same.

To add $\frac{2x+5}{4x-3}+\frac{x-7}{4x-3}$, we'll add $(2x+5)+(x-7)$ to form the new numerator, but the denominator will remain $4x-3$.

$$\frac{2x+5}{4x-3}+\frac{x-7}{4x-3}=\frac{(2x+5)+(x-7)}{4x-3}=\frac{(2x+x)+(5-7)}{4x-3}=\frac{3x-2}{4x-3}$$

 TIP

Always check to see if the sum (or difference) can be simplified. Even if both fractions were in simplest form when you started, you may be able to simplify the result.

All the usual rules apply. Combine like terms and only like terms. Simplify whenever possible, but cancel factors, not terms. That means that after you add (or subtract), you may need to factor the numerator and/or the denominator to see if there's anything to cancel. It won't happen every time, but it does happen. Here's an example.

$$\frac{x^2+5x-19}{x+2}+\frac{15-5x}{x+2}=\frac{x^2+5x-5x-19+15}{x+2}=\frac{x^2-4}{x+2}=\frac{(x+2)(x-2)}{x+2}=x-2$$

For subtraction, there is another concern to keep in mind. Remember that the fraction bar acts like a set of parentheses around the numerator, so when you place a subtraction sign between two fractions, it's important to remember that you must subtract the entire second numerator.

When asked to subtract $\frac{2x^2+6x-5}{2x+1}-\frac{x^2+4x-7}{2x+1}$, you have to subtract x^2, subtract $4x$, and subtract -7. The simplest way to remember that is to put in parentheses and write the problem as $\frac{(2x^2+6x-5)-(x^2+4x-7)}{2x+1}$. Distribute the minus by changing all the signs of the second polynomial, and then you can add.

$$\frac{2x^2+6x-5}{2x+1}-\frac{x^2+4x-7}{2x+1}=\frac{\left(2x^2+6x-5\right)-\left(x^2+4x-7\right)}{2x+1}$$

$$=\frac{2x^2+6x-5-x^2-4x+7}{2x+1}$$

$$=\frac{2x^2-x^2+6x-4x-5+7}{2x+1}$$

$$=\frac{x^2+2x+2}{2x+1}$$

Through all of that, our denominator stays the same.

CHECK POINT

Add or subtract, and simplify if possible.

31. $\dfrac{t}{t+5}+\dfrac{5}{t+5}$

32. $\dfrac{5x+1}{x+3}+\dfrac{5-3x}{x+3}$

33. $\dfrac{5x}{x^2-4}-\dfrac{10}{x^2-4}$

34. $\dfrac{3x+2}{7x}+\dfrac{5x+7}{7x}$

35. $\dfrac{2y-3}{9y}-\dfrac{5y-4}{9y}$

36. $\dfrac{12x-5}{2x+5}-\dfrac{8x-5}{2x+5}$

37. $\dfrac{x^2-7}{x-y}-\dfrac{y^2-7}{x-y}$

38. $\dfrac{2x^2+9x+4}{x^2-5x+6}-\dfrac{x^2+15x-5}{x^2-5x+6}$

39. $\dfrac{3x-2}{5x}+\dfrac{2x-3}{5x}-\dfrac{5-2x}{5x}$

40. $\dfrac{2x+19}{x^2+3x+2}+\dfrac{7x-16}{x^2+3x+2}-\dfrac{3x-9}{x^2+3x+2}$

Addition and Subtraction with Unlike Denominators

When you need to add or subtract fractions with different denominators, you can't plunge in and add the numerators. That would be like trying to combine apples and oranges. Different denominators let you know that you're dealing with different kinds of things. If you have 5 apples and 3 oranges, you don't have 8 apples or 8 oranges. You have 8 fruits. You have to find a description that fits both apples and oranges, a common denominator.

To find the common denominator for fractions, we find the lowest common multiple of the two denominators, but when we're working with rational expressions, the common denominator isn't immediately apparent. You can find a common denominator for $\dfrac{3}{5}$ and $\dfrac{1}{4}$ without too much work, but finding a common denominator for $\dfrac{x+5}{x^2-6x+8}$ and $\dfrac{3x}{x^2-16}$ is more of a challenge. So let's take this step by step.

1. If any denominators can be factored, factor them. Do not factor numerators. Denominators may or may not factor, and some may factor and some not.

2. Multiply each denominator by any factors present in other denominators, but not already present in this denominator. Do not simplify the denominators. Leave them in factored form.

3. Multiply each numerator by the same factors by which you multiplied its denominator. Simplify the numerators.

4. For subtraction, change the signs of the second numerator and add.

5. Add the numerators by combining like terms.

6. Check to see if the numerator of the sum will factor. Specifically, check to see if any of the factors of the common denominator are factors of the numerator. Cancel if possible.

Let's look at an example in which the denominators do factor. We'll subtract $\dfrac{5x+2}{x^2-4x+3}-\dfrac{3x-4}{x^2-8x+15}$.

1. If any denominators can be factored, factor them.

$$\frac{5x+2}{x^2-4x+3}-\frac{3x-4}{x^2-8x+15}$$
$$\frac{5x+2}{(x-3)(x-1)}-\frac{3x-4}{(x-3)(x-5)}$$

2. Multiply each denominator by any factors present in other denominators, but not already present in this denominator.

$$\frac{(5x+2)}{(x-3)(x-1)_{(x-5)}}-\frac{(3x-4)}{(x-3)(x-5)_{(x-1)}}$$

3. Multiply each numerator by the same factors.

$$\frac{(5x+2)_{(x-5)}}{(x-3)(x-1)_{(x-5)}}-\frac{(3x-4)_{(x-1)}}{(x-3)(x-5)_{(x-1)}}$$
$$\frac{5x^2-23x-10}{(x-3)(x-1)(x-5)}-\frac{3x^2-7x+4}{(x-3)(x-5)(x-1)}$$

4. For subtraction, change the signs and add.

$$\frac{5x^2-23x-10}{(x-3)(x-1)(x-5)}-\frac{3x^2-7x+4}{(x-3)(x-5)(x-1)}$$
$$\frac{5x^2-23x-10}{(x-3)(x-1)(x-5)}+\frac{-3x^2+7x-4}{(x-3)(x-5)(x-1)}$$

5. Add the numerators.

$$\frac{5x^2 - 23x - 10}{(x-3)(x-1)(x-5)} + \frac{-3x^2 + 7x - 4}{(x-3)(x-5)(x-1)}$$

$$\frac{5x^2 - 3x^2 - 23x + 7x - 10 - 4}{(x-3)(x-1)(x-5)}$$

$$\frac{2x^2 - 16x - 14}{(x-3)(x-1)(x-5)}$$

6. Check to see if the numerator will factor. Cancel if possible.

$$\frac{2(x^2 - 8x - 7)}{(x-3)(x-1)(x-5)}$$

You were able to factor out a common factor of 2, but nothing will cancel. You can present the result as $\dfrac{2(x^2 - 8x - 7)}{(x-3)(x-1)(x-5)}$ or $\dfrac{2x^2 - 16x - 14}{(x-3)(x-1)(x-5)}$, or with the denominator multiplied out as $\dfrac{2x^2 - 16x - 14}{x^3 - 9x^2 + 23x - 15}$, but that multiplication isn't usually valuable enough to make the work worthwhile.

Don't be concerned if the denominators don't factor. You can still find a common denominator. To add $\dfrac{x-3}{x+5} + \dfrac{x+1}{x-7}$, where the denominators don't factor, just think of each denominator as a one-factor denominator. To create a common denominator, follow the other remaining steps.

$$\frac{x-3}{(x+5)} + \frac{x+1}{(x-7)}$$

$$\frac{x-3}{(x+5)_{(x-7)}} + \frac{x+1}{(x-7)_{(x+5)}}$$

$$\frac{(x-3)_{(x-7)}}{(x+5)_{(x-7)}} + \frac{(x+1)_{(x+5)}}{(x-7)_{(x+5)}}$$

$$\frac{x^2 - 10x + 21}{(x+5)(x-7)} + \frac{x^2 + 6x + 5}{(x+5)(x-7)}$$

$$\frac{2x^2 - 4x + 26}{(x+5)(x-7)}$$

$$\frac{2(x^2 - 2x + 13)}{(x+5)(x-7)}$$

That numerator does not factor any more than that, and nothing cancels, so we can stop there. The common denominator is just the product of the two denominators. You could write the answer as $\dfrac{2x^2 - 4x + 26}{x^2 - 2x - 35}$ if you prefer.

CHECK POINT

Add or subtract, and simplify if possible.

41. $\dfrac{a}{a+3} + \dfrac{2a}{3(a+3)}$

42. $\dfrac{7}{3x+6} - \dfrac{2}{5x+10}$

43. $\dfrac{5}{7} - \dfrac{2}{x+7}$

44. $\dfrac{x}{x+5} + \dfrac{5}{x-5}$

45. $\dfrac{8}{x+6} + \dfrac{7}{x^2+6x}$

46. $\dfrac{x+2}{x+3} + \dfrac{x+3}{x+4}$

47. $\dfrac{y-7}{y-9} - \dfrac{y-2}{y-3}$

48. $\dfrac{6x}{x^2+5x+6} - \dfrac{2}{x+3}$

49. $\dfrac{4}{x^2+7x+10} - \dfrac{1}{x^2-25}$

50. $\dfrac{2a+3}{a^2+4a+4} - \dfrac{a-2}{a^2+5a+6}$

The Least You Need to Know

- A rational expression is the quotient of two polynomials, and is defined for all real numbers that do not make the denominator equal to zero.

- Simplify rational expressions by factoring the numerator and denominator and cancel any factor that appears in both.

- Multiply rational expressions by factoring all numerators and denominators, canceling factors from either numerator with matching factors in any denominator, and then multiplying the numerators and multiplying the denominators.

- Divide rational expressions by inverting the divisor and multiplying.

- If two rational expressions have the same denominator, add or subtract them by adding or subtracting the numerators and keeping the same denominator.

- When denominators are different, factor each denominator. Multiply the numerator and denominator of each rational expression by any factors from the other denominator that are lacking. Once you have the same denominators, add or subtract the numerators.

Rational Equations and Functions

Algebra begins with variables, which quickly combine into expressions and then equations and inequalities. We started by solving linear equations, and then explored equations that involved radicals and others that involved squares. Those quadratic equations were built from polynomials and our most recent bit of building was assembling polynomials into rational expressions. Now it's time to look at equations and functions built from rational expressions.

In this chapter, we'll investigate the two most common ways to solve equations involving rational expressions, and we'll explore the characteristics of the graphs of rational functions. As always, we'll look for shortcuts, and make good use of skills we've already built, like factoring and operations with polynomials.

In This Chapter

- Solving rational equations by cross-multiplying

- Multiplying through to clear denominators

- Graphing rational functions

Solving Rational Equations

An equation that contains one or more rational expressions is called a *rational equation*. When solving rational expressions, you need to be aware of the domain of the equation. Each rational expression has a domain, a set of all the real numbers that could be substituted for the variable without making the denominator equal zero. The expression $\dfrac{x+1}{x-3}$ has a domain of all real numbers except 3, and the expression $\dfrac{24}{x^2-1}$ has a domain of all real numbers except 1 and -1. If we build a rational equation that includes both of those expressions, such as $\dfrac{x+1}{x-3}+\dfrac{24}{x^2-1}=4$, the domain of the equation can't include 3, because 3 is not in the domain of $\dfrac{x+1}{x-3}$ and can't include 1 or -1, because they're not in the domain of $\dfrac{24}{x^2-1}$. The domain of the equation is all real numbers except -1, 1, and 3.

 **DEFINITION**

A **rational equation** is an equation that contains one or more rational expressions.

You need to be aware of the domain of a rational equation whenever you set out to find its solution. Rational equations may have more than one solution, or one solution, or no solutions. Whenever you solve rational equations, however, you have to be alert to the possibility of an extraneous solution, and checking your solutions will be crucial.

A rational equation may be as simple as $\dfrac{1}{x}=\dfrac{1}{2}$, which you can probably guess has a solution of $x=2$, but they can get quite complicated. We'll look at two methods for solving them, and what conditions have to be met before you can use each one. Rational expressions take their name from ratio, a comparison by division. Two equal ratios create an equation called a proportion, and the principal strategy for finding a missing term of a proportion is cross-multiplying.

Cross-Multiplying

In the simple rational equation $\dfrac{1}{x}=\dfrac{1}{2}$, you could probably determine that $x=2$ by just visually matching the two sides. A rational equation that has one rational expression equal to a constant, or equal to another single rational expression, is basically a proportion.

In any proportion, the product of the means is equal to the product of the extremes, which means that if $\dfrac{1}{x}=\dfrac{1}{2}$, then $x\cdot1=1\cdot2$. In this simple equation, that leads directly to $x=2$, but even in more complicated equations, if the structure of the equation is two equal ratios, you can cross-multiply and solve the resulting equation.

Let's look at some examples that are not quite so simple. Let's start by solving $\dfrac{x+2}{x+5}=\dfrac{1}{2}$. You have one rational expression equal to a constant, so you can cross-multiply. Before doing that, think for a moment about the domain. Because the denominator is $x+5$, the domain cannot include –5. The domain is all real numbers except –5. Let's solve.

$$\frac{x+2}{x+5}=\frac{1}{2}$$
$$1(x+5)=2(x+2)$$
$$x+5=2x+4$$
$$5=x+4$$
$$1=x$$

A solution of $x=1$ is possible because 1 is in the domain, but you should still check to be sure that it is a valid solution. Check by replacing each x in the equation with 1.

$$\frac{x+2}{x+5}=\frac{1}{2}\Rightarrow\frac{1+2}{1+5}=\frac{1}{2}\Rightarrow\frac{3}{6}=\frac{1}{2}$$

That substitution gives a true statement, so the solution is correct.

Let's try one that leads to a quadratic equation. Before solving $\frac{x-3}{2x+3}=\frac{x-5}{x-1}$, notice that the domain is all real numbers except 1 and $-\frac{3}{2}$. If either of those values turn up as solutions, we'll reject them as extraneous. Let's cross-multiply and see where it takes us.

$$\frac{x-3}{2x+3}=\frac{x-5}{x-1}$$
$$(2x+3)(x-5)=(x-3)(x-1)$$
$$2x^2-7x-15=x^2-4x+3$$

Cross-multiplying is going to leave you with a quadratic equation, so gather up all the terms on one side equal to zero.

$$x^2-7x-15=-4x+3$$
$$x^2-3x-15=3$$
$$x^2-3x-18=0$$

You can use the quadratic formula if necessary, but it's worthwhile to try factoring first.

$$x^2-3x-18=0$$
$$(x-6)(x+3)=0$$
$$x-6=0 \qquad x+3=0$$
$$x=6 \qquad x=-3$$

You end up with two solutions. Both are in the domain, but both need to be checked.

$$\frac{x-3}{2x+3}=\frac{x-5}{x-1}$$

$$\frac{6-3}{2\cdot6+3}=\frac{6-5}{6-1}$$

$$\frac{3}{15}=\frac{1}{5}$$

$$\frac{x-3}{2x+3}=\frac{x-5}{x-1}$$

$$\frac{-3-3}{2(-3)+3}=\frac{-3-5}{-3-1}$$

$$\frac{-6}{-3}=\frac{-8}{-4}$$

 ALGEBRA TRAP

Forgetting to check the solutions to a rational equation is one of the most common sources of errors. Extraneous solutions are possible, especially when there are values that are excluded from the domain. Always check rational equations by returning to the original equation so you don't get confused by any errors in your work.

Both solutions check in the original equation, so the rational equation $\frac{x-3}{2x+3}=\frac{x-5}{x-1}$ has two solutions, $x=6$ and $x=-3$.

Cross-multiplying is only possible when you have one rational expression equal to one rational expression (or to a constant), and that may not sound like it covers many rational equations. It doesn't seem to work for an equation like $\frac{x+1}{x-4}+\frac{24}{x-1}=5$.

Don't forget the skills you already have, however. The equation $\frac{x+1}{x-4}+\frac{24}{x-1}=5$ has a sum of rational expressions equal to a constant, but you know how to add rational expressions. You could add those two expressions to create one rational expression that is equal to the constant 5.

$$\frac{x+1}{x-4}+\frac{24}{x-1}=5$$

$$\frac{(x+1)(x-1)}{(x-4)(x-1)}+\frac{24(x-4)}{(x-1)(x-4)}=5$$

$$\frac{x^2-1}{(x-4)(x-1)}+\frac{24x-96}{(x-1)(x-4)}=5$$

$$\frac{x^2+24x-97}{(x-4)(x-1)}=5$$

Now you have one rational expression equal to a constant. Multiply out the denominator, put the 5 over 1, and cross-multiply.

$$\frac{x^2 + 24x - 97}{(x-4)(x-1)} = 5$$

$$\frac{x^2 + 24x - 97}{x^2 - 5x + 4} = \frac{5}{1}$$

$$5(x^2 - 5x + 4) = 1(x^2 + 24x - 97)$$

$$5x^2 - 25x + 20 = x^2 + 24x - 97$$

Now you have a quadratic equation, so gather all terms on one side equal to zero.

$$5x^2 - 25x + 20 = x^2 + 24x - 97$$

$$4x^2 - 25x + 20 = 24x - 97$$

$$4x^2 - 49x + 20 = -97$$

$$4x^2 - 49x + 117 = 0$$

The size of the coefficients may make you think you need the quadratic formula, and you certainly can use it, but those numbers will be even bigger. It turns out that 117 can be factored as $1 \cdot 117$, $3 \cdot 39$, or $9 \cdot 13$, and that's not all that many possibilities to try.

$$4x^2 - 49x + 117 = 0$$

$$(4x - 13)(x - 9) = 0$$

$$4x - 13 = 0 \qquad x - 9 = 0$$

$$4x = 13 \qquad\quad x = 9$$

$$x = \frac{13}{4}$$

The domain of the equation is all real numbers except 1 and 4, so these solutions are acceptable, but you still should check them back in the original equation.

$$\frac{x+1}{x-4}+\frac{24}{x-1}=5$$

$$\frac{13/4+1}{13/4-4}+\frac{24}{13/4-1}=5$$

$$\frac{17/4}{-3/4}+\frac{24}{9/4}=5$$

$$-\frac{17}{3}+\frac{96}{9}=5$$

$$-\frac{17}{3}+\frac{32}{3}=5$$

$$\frac{15}{3}=5$$

$$\frac{x+1}{x-4}+\frac{24}{x-1}=5$$

$$\frac{9+1}{9-4}+\frac{24}{9-1}=5$$

$$\frac{10}{5}+\frac{24}{8}=5$$

$$2+3=5$$

 TIP

When you have fractions inside of fractions you can simplify by multiplying the numerator and denominator of the outer fraction by the denominator of the inner fraction. This will clear the denominators of the inner fractions.

That first check requires patience, but both solutions check. That's a lot of algebra to solve one equation, even if that equation does have two solutions. Next we'll look at a technique that may be quicker.

 CHECK POINT

Solve each equation by cross-multiplying. Be sure to check for extraneous solutions.

1. $\dfrac{8}{3x}=2$

2. $\dfrac{3-5x}{6}=3$

3. $\dfrac{4x-6}{x}=2$

4. $\dfrac{a}{a-5}=\dfrac{8}{3}$

5. $\dfrac{7x-4}{6x+4}=\dfrac{6}{7}$

6. $\dfrac{3}{5x-1}=\dfrac{2}{3x}$

7. $\dfrac{1}{x}=\dfrac{1}{x^2-2x}$

8. $\dfrac{20}{3x^2-6x}=\dfrac{5}{x}$

9. $\dfrac{5}{2x}=\dfrac{7}{2x-8}$

10. $\dfrac{-1}{x-2}=\dfrac{2}{x^2-4}$

Multiplying Through

Cross-multiplying is a convenient tactic for solving rational equations that involve only one rational expression, or equations that state two rational expressions are equal. When more operations are involved, however, simplifying the equation enough to use cross-multiplication can be time-consuming. Our second method often proves more practical in those cases.

Let's revisit $\frac{x+1}{x-4} + \frac{24}{x-1} = 5$, the equation you solved earlier. To use cross-multiplying, you had to add the rational expressions first. Then you cross-multiplied, which resulted in a quadratic equation that you had to solve and check. This time, you'll use a method called *multiplying through* to turn the rational equation into a simpler equation quickly. Here's the plan.

1. Find the common denominator for all the rational equations in the equation.

2. Multiply both sides of the equation by that common denominator, making sure to distribute where necessary. This should eliminate all denominators.

3. Solve the resulting equation.

4. Check any solutions in the original rational equation.

To try this with $\frac{x+1}{x-4} + \frac{24}{x-1} = 5$, you first need the common denominator. The common denominator needs to include a factor of $x-4$ and a factor of $x-1$, and you don't need to actually multiply it out. You can leave it in factored form, so the common denominator is $(x-4)(x-1)$.

Now multiply both sides of the equation $\frac{x+1}{x-4} + \frac{24}{x-1} = 5$ by $(x-4)(x-1)$.

$$\frac{x+1}{x-4} + \frac{24}{x-1} = 5$$

$$\left[(x-4)(x-1)\right]\left[\frac{x+1}{x-4} + \frac{24}{x-1}\right] = 5\left[(x-4)(x-1)\right]$$

It looks rather messy right now, but stay calm and be patient. Distribute on the left side and do some canceling.

$$\left[(x-4)(x-1)\right]\left[\frac{x+1}{x-4} + \frac{24}{x-1}\right] = 5\left[(x-4)(x-1)\right]$$

$$\left[\cancel{(x-4)}(x-1)\right]\left[\frac{x+1}{\cancel{x-4}}\right] + \left[(x-4)\cancel{(x-1)}\right]\left[\frac{24}{\cancel{x-1}}\right] = 5\left[(x-4)(x-1)\right]$$

This leaves you with a more manageable equation.

$$(x-1)(x+1)+(x-4)\cdot 24 = 5\big[(x-4)(x-1)\big]$$

The denominators are gone, so multiply out.

$$(x-1)(x+1)+(x-4)\cdot 24 = 5\big[(x-4)(x-1)\big]$$
$$(x^2-1)+(24x-96) = 5(x^2-5x+4)$$
$$x^2+24x-97 = 5x^2-25x+20$$

You've seen this equation before, when you solved it by cross-multiplying, and there's no need for you to solve it again, but getting to this point was quicker and easier with the multiplying through method.

 TIP

If you multiply through and still have a denominator, check to make sure you've used the correct common denominator, and that you've distributed to every term.

Let's try another example by multiplying through, and take this one all the way to the end. Suppose you want to solve $\dfrac{1}{x-2}+\dfrac{3}{x+3}=\dfrac{5}{x^2+x-6}$. Before looking for the common denominator, factor the last denominator. $x^2+x-6=(x+3)(x-2)$, so your equation is $\dfrac{1}{x-2}+\dfrac{3}{x+3}=\dfrac{5}{(x+3)(x-2)}$, and the common denominator for all three rational expressions is $(x+3)(x-2)$.

Multiply both sides of the equation by $(x+3)(x-2)$, distributing on the left side.

$$\big[(x+3)(x-2)\big]\left[\frac{1}{x-2}+\frac{3}{x+3}\right]=\left[\frac{5}{(x+3)(x-2)}\right]\big[(x+3)(x-2)\big]$$

$$\big[(x+3)(x-2)\big]\cdot\frac{1}{x-2}+\big[(x+3)(x-2)\big]\cdot\frac{3}{x+3}=\left[\frac{5}{(x+3)(x-2)}\right]\big[(x+3)(x-2)\big]$$

Canceling what you can will eliminate all the denominators.

$$\big[(x+3)\cancel{(x-2)}\big]\cdot\frac{1}{\cancel{x-2}}+\big[\cancel{(x+3)}(x-2)\big]\cdot\frac{3}{\cancel{x+3}}=\left[\frac{5}{\cancel{(x+3)(x-2)}}\right]\big[\cancel{(x+3)(x-2)}\big]$$

$$(x+3)+3(x-2)=5$$

This is a much simpler equation now, so let's solve. Simplify the left side first.

$$(x+3)+3(x-2)=5$$
$$x+3+3x-6=5$$
$$4x-3=5$$
$$4x=8$$
$$x=2$$

Don't forget to check the solution. The values that must be excluded from the domain are 2 and -3, so $x=2$ is not an acceptable solution.

$$\frac{1}{2-2}+\frac{3}{2+3}=\frac{5}{2^2+2-6}$$
$$\frac{1}{0}+\frac{3}{5}=\frac{5}{0}$$

Substituting $x=2$ in the original equation will make two of the three denominators equal zero. It cannot be a solution, even though all the algebra was done correctly. $x=2$ is an extraneous solution and the equation $\dfrac{1}{x-2}+\dfrac{3}{x+3}=\dfrac{5}{x^2+x-6}$ has no solution.

CHECK POINT

Solve each equation by multiplying through. Be sure to check for extraneous solutions.

11. $\dfrac{1}{x}+\dfrac{1}{3x}=28$

12. $\dfrac{5}{x}+\dfrac{15}{2x}=\dfrac{5}{4}$

13. $\dfrac{x+3}{x}-\dfrac{16}{15}=\dfrac{x-3}{3x}$

14. $\dfrac{x-3}{x+1}+\dfrac{1}{x-1}=1$

15. $\dfrac{3}{x-5}-\dfrac{17}{4x-20}=\dfrac{5}{8}$

16. $\dfrac{2}{t+4}+\dfrac{3}{t-4}=\dfrac{29}{t^2-16}$

17. $\dfrac{1}{y-5}-\dfrac{5}{3y+15}=\dfrac{8}{y^2-25}$

18. $\dfrac{2}{x-1}-\dfrac{4}{x^2-1}=\dfrac{3}{x+1}$

19. $\dfrac{5}{x+5}-\dfrac{3}{x+8}=\dfrac{7}{x^2+13x+40}$

20. $\dfrac{3x+13}{x^2-x-6}=\dfrac{4}{x-3}+\dfrac{1}{x+2}$

Graphing Simple Rational Functions

You've explored the graphs of several different kinds of functions. Now it's time to look at the graphs of functions that involve rational expressions, which are quite distinctive.

Because the domain cannot include any values of the variable that will make a denominator equal to zero, the domain of a rational function will usually be limited. Sometimes only one value has to be excluded, sometimes more than one, but even one excluded value will cause a break in the graph. This means that the graph of a rational function is usually in two or more pieces. Rational functions that have more than one value excluded from their domain will have graphs with more pieces. If a rational function is defined for all real numbers, its graph will be one piece.

A rational function for which only one value is excluded from the domain is called a simple rational function. The shape of the graph of a simple rational function is a pair of curves that together are called a hyperbola. They look like a pair of wings turning away from one another.

Let's begin our look at graphs of rational functions with a simple case, the graph of $y = \frac{8}{x}$. When you graphed parabolas, you learned that to build a good picture of the function, you needed to locate the vertex before setting up our table of values. You found at least the x-coordinate of the vertex so that you could be sure to include a few points to the right and a few points to the left of the vertex.

For rational functions, you want to locate the value that makes the denominator equal to zero. That is the x-value where the function is undefined and therefore where there will be a break in the graph. It's called a *discontinuity*. You don't want to include that value, but you want to make sure that you choose some x-values below the discontinuity and some above it, so that you see both pieces of the graph.

 **DEFINITION**

A **discontinuity** is the value at which the rational function is undefined, because there is a break in the graph at that value.

For $y = \frac{8}{x}$, the value that must be excluded from the domain is $x = 0$, so we'll choose some negative values and some positive values for x. We'll have to calculate the quotient of 8 and our x-value, so we'll start with factors of 8.

x	-8	-4	-2	-1	1	2	4	8
y	-1	-2	-4	-8	8	4	2	1
(x, y)	(-8, -1)	(-4, -2)	(-2, -4)	(-1, -8)	(1, 8)	(2, 4)	(4, 2)	(8 , 1)

Here's what it looks like when you plot the points. Can you see the two wings?

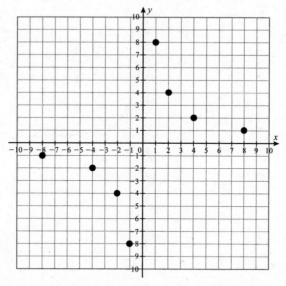

Before trying to connect the dots, there are a couple of key characteristics of the graph to note. You know that there would be a discontinuity, a break in the graph, at $x = 0$, so it's important to remember not to try to connect the two segments of the graph. Each wing is a curve, as you may already be able to tell from the points, so put the ruler away. Sketch the curves smoothly. Finally, let's look at what happens as x becomes very large or very small.

The simplest way to see what happens is to evaluate the function for a few large values of x, and a few small ones.

x	10	100	1,000	-10	-100	-1,000
y	0.8	0.08	0.008	-0.8	-0.08	-0.008

As x becomes very large, the value of y gets smaller and smaller, coming closer to 0, but as long as x is a positive number, $y = \dfrac{8}{x}$ will be a positive number. There is no value of x that will ever make y equal zero, so the right wing of the graph will get very close to the x-axis, but will never cross it. In the same way, the left wing will always have negative y-values, getting closer and closer to the x-axis but never touching it.

When you sketch the graph with those concerns in mind, it looks like this.

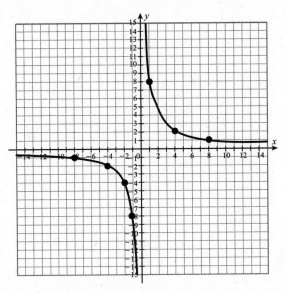

Let's investigate another example, and this time we'll make the equation a little more complicated, but just a little. Let's graph $y = \dfrac{2x+2}{x-3}$.

Start by identifying the domain. The denominator is $x - 3$ so $x = 3$ would make the denominator zero. The domain is all real numbers except $x = 3$, and your graph will have a discontinuity at $x = 3$.

Choose a few values less than 3 and a few greater than 3 to build a table of values.

x	-1	0	1	2	4	5	6	7
$2x+2$	0	2	4	6	10	12	14	16
$x-3$	-4	-3	-2	-1	1	2	3	4
y	0	$-\dfrac{2}{3}$	-2	-6	10	6	$4\dfrac{2}{3}$	4
(x, y)	(-1, 0)	$\left(0, -\dfrac{2}{3}\right)$	(1, -2)	(2, -6)	(4, 10)	(5, 6)	$\left(6, 4\dfrac{2}{3}\right)$	(7, 4)

The shape of the graph is a little harder to spot from these points. The first graph we looked at was compartmentalized by the x-axis and y-axis. Because the graph never crossed those axes, each wing was contained in one quadrant. Here the boundaries aren't as clear, but if you stop to think for a moment, you can find them.

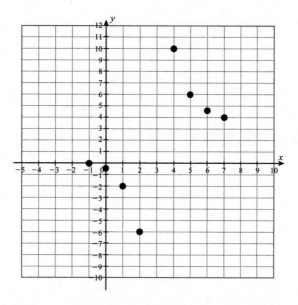

You know that $x = 3$ is not in the domain, so there will be a discontinuity there. Draw in the vertical line $x = 3$ as a dotted line on the graph.

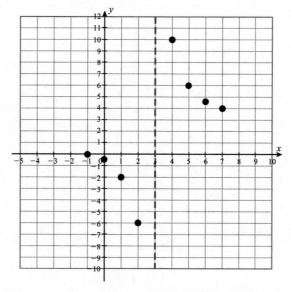

That divides the two wings, the way the y-axis did in our first example. That first example also had the x-axis as a dividing line between the two wings, with both getting close to the x-axis but never touching it. That's not the case here. You already have a point on the x-axis. Let me show you a few more points that fit our equation. I think you'll be able to guess where the dividing line is.

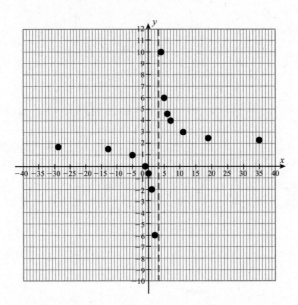

Can you see that $y = 2$ is the dividing line? If you add a dotted line there, and sketch in the graph, it looks like this.

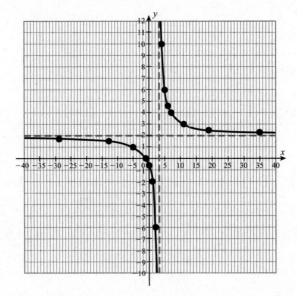

Asymptotes

Those two dividing lines, the vertical line and the horizontal lines that the graph gets very close to but doesn't touch, are called *asymptotes*. Although they're not part of the graph, knowing where they are helps us draw the graph of the rational function. Because the asymptotes are helpful, having a quick way to find them would make the task of graphing a rational function easier.

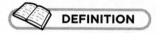 **DEFINITION**

An **asymptote** is a line that a graph approaches very closely. The graph never crosses a vertical asymptote.

The vertical asymptote is easy to locate, and in fact, you find it every time you start to graph a radical function. The first thing you do is find the value that must be excluded from the domain, the location of the discontinuity, and that's where the vertical asymptote is. If you find that $x = -4$ makes your denominator zero, then x cannot equal -4, so there's a break in the graph there. The discontinuity occurs at -4, and the equation of the vertical asymptote is $x = -4$.

The horizontal asymptote takes a little more work, but for simple rational functions, you can find it with one of two rules.

1. If the degree of the numerator and the degree of the denominator are the same, as they are in $y = \dfrac{2x+2}{x-3}$, divide the highest power term of the numerator by the highest power term of the denominator. In our example $\dfrac{2x}{x} = 2$, so our horizontal asymptote is $y = 2$.

2. If the degree of the numerator is less than the degree of the denominator, as it is in $y = \dfrac{8}{x}$, the horizontal asymptote is $y = 0$, the x-axis.

You might be wondering if there's a third rule for when the degree of the numerator is higher than the degree of the denominator. You could call it a rule, I guess, but if that happens, there's no horizontal asymptote. That won't occur in simple rational equations.

Intercepts

As you saw when you graphed $y = \dfrac{8}{x}$, not every rational function will have an x-intercept or a y-intercept, but when there are intercepts, they're easy to find and helpful for graphing. By replacing x with 0 and simplifying, you can find the y-intercept. When there is no y-intercept, you'll have an expression like $y = \dfrac{8}{0}$, which is undefined.

To find the x-intercepts, replace y with 0 and solve the equation. For a simple rational function like $y = \dfrac{2x+2}{x-3}$, doing that will give you an equation you'll want to solve by cross-multiplying.

$$\frac{0}{1} = \frac{2x+2}{x-3}$$
$$1(2x+2) = 0(x-3)$$
$$2x+2 = 0$$

Solving a simple rational function to find the x-intercepts turns into finding the values of x that make the numerator equal zero.

THINK ABOUT IT

Replacing y with 0 and solving the resulting rational equation will turn into solving for values that make the numerator zero. For a rational expression, a fraction, to equal zero, the fraction must be zero over a non-zero number. No other quotient can equal zero.

Step-by-step, here's the plan for graphing simple rational functions.

1. Find the value of x that will make the denominator equal zero. This value is not in the domain, but it is the location of the vertical asymptote. Draw a dotted vertical line for the vertical asymptote.

2. Find the horizontal asymptote by following our two rules ($y = 0$ or $y = \dfrac{\text{lead term}}{\text{lead term}}$). Draw a dotted horizontal line for the horizontal asymptote.

3. The two asymptotes divide the plane into four sections. The two wings of the hyperbola will be in two of the four.

4. Find the y-intercept if there is one, by substituting 0 for x and simplifying. Plot the y-intercept.

5. Find the x-intercept, if there is one, by finding the values of x that make the numerator equal to zero. Plot the x-intercept.

6. Make a table of values, using as many values as necessary to define the curves. Be sure to choose values on both sides of the vertical asymptote.

7. Draw each wing by connecting the points with a smooth curve. Do not connect the two wings. Do not cross the asymptotes.

TIP

For simple rational functions in which x-terms are first degree, the two wings will be in diagonally opposite sections. The wings will be either in the upper right and lower left sections, or in the lower right and upper left. But when higher powers of x are involved, you may find the wings in side-by-side sections.

CHECK POINT

Graph each rational function by finding asymptotes and intercepts, and making a table of values if necessary.

21. $y = \dfrac{4}{x}$

22. $y = \dfrac{-12}{x}$

23. $y = \dfrac{3}{x-2}$

24. $y = \dfrac{x}{x+4}$

25. $y = \dfrac{-6}{x-3}$

26. $y = \dfrac{x-1}{x+5}$

27. $y = \dfrac{x+3}{x-4}$

28. $y = \dfrac{2x+6}{x-1}$

29. $y = \dfrac{x+5}{2x}$

30. $y = \dfrac{2x}{x+5}$

The Least You Need to Know

- Solve rational equations that are two equal rational expressions by cross-multiplying. Solve the resulting equation and be sure to check your solution.
- Solve multi-term rational equations by multiplying through by the common denominator of all rational expressions involved. Solve the resulting equation and check your solution.
- Simple rational functions have a graph called a hyperbola.
- The vertical asymptote(s) occurs at the point(s) of discontinuity. The horizontal asymptote occurs at the ratio of the lead terms, or at the x-axis.
- Sketch the graph of the rational function by finding the vertical and horizontal asymptotes, the intercepts if any, and then building a table.

Glossary

absolute value The distance of a number from zero, without regard to direction. The size or magnitude of a number without regard to its sign.

addend Each of the numbers that are added in an addition problem.

algorithm A list of steps necessary to perform a process.

associative property A property of addition or multiplication that says that when adding or multiplying more than two numbers, you may group them in different ways without changing the result.

asymptote A line that a graph approaches very closely. Graphs never cross vertical asymptotes but may sometimes cross horizontal asymptotes.

axis A vertical or horizontal line that divides the coordinate plane into sections. The horizontal line is called the *x*-axis and the vertical line called the *y*-axis.

axis of symmetry An imaginary line passing through the vertex of a parabola, which divides the parabola into two sections that are reflections of one another.

base When an exponent is used to show repeated multiplication, the number to be multiplied is the base of the power.

binary operation A process that works on two numbers at a time.

binary system A place value system based on the number 2.

canceling The process of simplifying a multiplication of fractions by dividing a numerator and a denominator by a common factor.

Cartesian plane   A system that identifies every point in the plane by an ordered pair of numbers. Also called the coordinate plane.

closure   A property that says that adding or multiplying two numbers of a set results in another number in the set.

coefficient   The numerical part of a term. Technically, a term is made up of a numerical coefficient and a variable coefficient, but usually the variable coefficient is just called the variable and the numerical coefficient is called the coefficient.

common denominator   A multiple of the denominators of two or more fractions.

common fraction   Fractions written as a quotient of two integers.

commutative property   A property of addition or multiplication that says that reversing the order of the two numbers will not change the result.

compound inequality   Two inequalities connected with a conjunction, either the word *and* or the word *or*.

conjugate   The conjugate of an expression that is the sum of two terms is an expression that is the difference of those two terms. $a-b$ is the conjugate of $a+b$, and $a+b$ is the conjugate of $a-b$.

conjunction   A compound inequality in which the word *and* is used.

constants   Terms containing only numbers and no variables.

coordinate system   A system that locates every point in the plane by an ordered pair of numbers, (x, y). Also called the Cartesian plane.

counting numbers   The set of numbers $\{1, 2, 3, 4, ...\}$ They are the numbers used to count. The counting numbers are also called the natural numbers.

cross-multiply   To find the product of the means and the product of the extremes of a proportion, and to say that those products are equal.

decimal fraction   Fractions written in the base ten system with digits to the right of the decimal point.

decimal system   A place value system in which each position in which a digit can be placed is worth ten times as much as the place to its right.

degree   The degree of a monomial with one variable is the power of the variable.

denominator   The number below the division bar in a fraction that tells how many parts the whole was broken into, or what kind of fraction you have.

difference   The result of a subtraction problem.

digit   A single symbol that tells how many.

discontinuity The value at which the rational function is undefined is called a discontinuity, because there is a break in the graph at that value.

discriminant The discriminant is the portion of the quadratic formula that appears under the radical, $b^2 - 4ac$. It can be used to tell how many solutions (and what type of solutions) a quadratic equation will produce.

disjunction A compound inequality in which the two inequalities are joined by *or*.

distributive property A property that says that for any three numbers a, b, and c, $c(a + b) = ca + cb$. The answer you get by first adding a and b and then multiplying the sum by c will be the same as the answer you get by multiplying a by c and b by c and then adding the results.

dividend In a division problem, the number that is divided.

divisor In a division problem, the number by which you divide.

domain The set of all values that can be substituted in a function for x, that is, all possible inputs.

domain, rational expression The set of all real numbers that can be substituted in a rational expression for the variable without making the denominator equal to zero.

double root A solution that occurs twice for the same equation, because the equation was a perfect square trinomial. Also called a double solution.

equation A mathematical sentence, which often contains a variable.

exponent The small raised number that tells how many times to use the base number as a factor.

exponentiation The operation of raising to a power.

expression Any mathematical calculation.

extended ratio Several related ratios condensed into one statement. The ratios $a{:}b$, $b{:}c$, and $a{:}c$ make the extended ratio $a{:}b{:}c$.

extraneous solution A solution produced by correct algebraic procedures that does not satisfy the equation.

extremes The first and last numbers of a proportion.

factor Each of the numbers or variables being multiplied.

factor tree A method of finding the prime factorization of a number by starting with one factor pair and then factoring each of those factors, continuing until no possible factoring remains.

factoring The process of rewriting an integer as the product of two integers, or a polynomial as the product of two polynomials of lesser degree.

FOIL An acronym for First, Outer, Inner, Last. It summarizes the four multiplications necessary when multiplying two binomials.

fraction A symbol that represents part of a whole.

function A relation in which each number in the domain has only one partner from the range.

function, linear A function of the form $y = mx + b$. It defines the output variable as a multiple of the input variable, possibly plus or minus a constant.

greatest common factor The greatest common factor of two numbers is the largest number that is a factor of both.

identity A property that says that adding 0 or multiplying by 1 leaves a number unchanged.

imaginary numbers Numbers that do not fit in the real number system, but that mathematicians use to solve certain types of problems.

improper fraction A fraction whose value is more than 1. The numerator is larger than the denominator.

index A small number appearing in the crook of the radical sign that tells what power is being undone. If no index appears, the radical indicates a square root.

inequality A sentence that compares two expressions that are not equal and shows which one is larger.

inequaltity, compound Two inequalities connected with a conjunction, either the word *and* or the word *or*.

inequality, linear A statement that defines a relation in which the multiple outputs are less than, or greater than, or equal to, an expression involving the input value.

input The numbers from the domain that are put into a function.

integers The set of numbers that includes all the positive whole numbers, and their opposites, the negative whole numbers, and zero.

interest Money you pay for the use of money you borrow, or money you receive because you've put your money into a bank account or other investment.

interest rate The percent of the principal that will be paid in interest each year.

inverse A property that says that every operation can be reversed. Every number has an opposite and adding the opposite to a number brings you back to 0. Every number except 0 has a reciprocal and multiplying by the reciprocal gets back to 1. Zero is its own opposite but has no reciprocal.

inverse operation An operation that reverses the work of another.

irrational numbers Numbers that cannot be written as the quotient of two integers.

least common denominator The least common multiple of two or more denominators.

least common multiple The smallest number that has each of two or more numbers as a factor.

like terms Terms that have the same variable, raised to the same power.

line A set of points that has length but no width or height.

linear function A function of the form $y = mx + b$. It defines the value of the output variable as a multiple of the input variable, possibly plus or minus a constant.

linear inequality A statement that defines a relation (not a function) in which the output is less than, or greater than, or less than or equal to, or greater than or equal to, some expression involving the input value.

linear system A set of two linear equations, each involving the same two variables. The solution of the system is the one pair of values that satisfy both equations.

mean The arithmetic average of a group of numbers, found by adding all the numbers and dividing by the number of numbers in the group.

means of a proportion The two middle numbers in a proportion.

minuend The first number in a subtraction problem.

mixed number A whole number and a fraction, written side by side, representing the whole number plus the fraction.

monomial A monomial is a constant, a variable, or a product of constants and variables.

natural numbers The set of numbers {1, 2, 3, 4, …} They are the numbers you use to count. Also called the counting numbers.

negative exponent A negative exponent on a non-zero base represents a fraction with a numerator of 1 and a denominator of the base raised to the corresponding positive power.

number line A line divided into segments of equal length, labeled with numbers, usually the integers. Positive numbers increase to the right of zero, and negative numbers go down to the left.

numerator The number above the bar in a fraction, which tells you how many of that denomination are present.

order of operations An agreement among mathematicians regarding the order in which operations are performed when solving an equation. Operations enclosed in parentheses or other grouping symbols are performed first, followed by evaluating exponents. After that, do multiplication and division as you meet them moving left to right, and finally do addition and subtraction as you meet them, moving left to right.

ordered pair Two numbers, usually designated as x and y, that locate a point in a coordinate system.

origin The point at which the *x*-axis and *y*-axis intersect in a coordinate plane.

output The numbers in the range produced by a function.

parabola A cup-shaped graph characteristic of quadratic functions. Technically, it is defined as a set of points, each of which is equidistant from a predetermined point and line.

parallel lines Lines on the same plane that never intersect.

partial product In a multiplication of polynomials, or in the multiplication of multi-digit numbers, the polynomial produced by multiplying by just one digit or term is a partial product.

PEMDAS A mnemonic, or memory device, to help you remember that the order of operations is parentheses, exponents, multiplication and division, addition and subtraction.

percent A ratio that compares numbers to 100. 42 percent means 42 out of 100, or 42:100.

perpendicular lines Lines that meet to form a right angle.

place value system A number system in which the value of a symbol depends on where it is placed in a string of symbols.

plane A flat surface that has length and width but no thickness.

point A position in space that has no length, width, or height.

polynomial A polynomial is a sum of monomials.

power A way to tell how many times a number should be used in repeated multiplication. The number to be multiplied is the base of the power, and the small raised number that tells how many times to use it is called the exponent.

power of ten A number formed by multiplying several 10s. The first power of ten is 10. The second power of ten is 100, and the third power of 10 is 1,000.

prime factorization The prime factorization of a number is a multiplication that uses only prime numbers and produces the original number as its product.

prime number A whole number whose only factors are itself and 1.

principal The amount of money borrowed or invested. The rate is the percent of the principal that will be paid in interest each year.

product The result of multiplying.

proper fraction A fraction whose value is less than one.

proportion Two equal ratios. The means of a proportion are the two middle numbers. The extremes are the first and last number.

quadrants The four sections into which the coordinate plane is divided by the axes.

quadratic An equation that includes a term in which the variable is squared. There may also be a variable term in which the variable appears but is not squared, and a constant term.

quadratic formula The formula $x = \dfrac{-b \pm \sqrt{b^2 - 4ac}}{2a}$, which can be used to find solutions for any quadratic equation of the form $ax^2 + bx + c = 0$ by substituting the values of a, b, and c and simplifying.

quotient The result of division.

radical The symbol for the square root. From the Latin word *radix*, meaning root.

radicand The number or expression under a radical.

range The set of all outputs, all values of y.

rate A comparison of two quantities in different units, for example, miles per hour or dollars per day.

ratio A comparison of two numbers by division.

rational equation An equation that contains one or more rational expressions.

rational expression A quotient of two polynomials, provided that the polynomial in the denominator is not zero.

rational numbers The set of all numbers that can be written as the quotient of two integers.

rationalizing the denominator The process of changing the appearance, but not the value, of a quotient so that no radicals remain in the denominator.

real numbers The name given to the set of all rational numbers and all irrational numbers.

reciprocal Two numbers are reciprocals if their product is 1. Each number is the reciprocal of the other.

relation A pairing of numbers from one set, called the domain, with numbers from another set, called the range.

relatively prime Two numbers are relatively prime if the only factor they have in common is 1.

remainder A remainder is the number left over at the end of a division problem. It's the difference between the dividend and the product of the divisor and quotient.

root The opposite operation of a power, a way of undoing a power.

slope The slope of a line is a number that compares the rise or fall of a line to its horizontal movement.

solving an equation An equation is a mathematical sentence, which often contains a variable. Solving an equation is a process of isolating the variable to find the value that can replace the variable to make a true statement.

square numbers Numbers created by raising a number to the second power.

standard form The standard form of a polynomial writes the terms in order from highest degree to the lowest.

subtrahend In a subtraction problem, the number that is taken away.

sum The result of addition.

system A set of two linear equations, each involving the same two variables. The solution of the system is the one pair of values that satisfy both equations.

term An algebraic expression made up of numbers, variables, or both, that are connected only by multiplication.

unlike terms Terms with different variables, such as x and y.

variable A letter or symbol that takes the place of a number.

vertex The turning point of a graph.

whole numbers The set of numbers $\{0, 1, 2, 3, 4, ...\}$ formed by adding a zero to the counting numbers.

x-coordinate The first number in an ordered pair, which indicates horizontal movement.

y-coordinate The second number in an ordered pair, which indicates vertical movement.

zero exponent Any non-zero number to the 0 power is 1.

Check Point Answers

Chapter 1

1. $-\dfrac{5}{13}$ is b) a rational number and d) a real number.

2. $\sqrt{49} = 7$ so it is a) an integer, b) a rational number, and d) a real number.

3. $\sqrt{5}$ is c) an irrational number and d) a real number.

4. $\sqrt{6.25} = 2.5$ so it is b) a rational number and d) a real number.

5. distributive property

6. inverse (for addition)

7. commutative property (for addition)

8. closure

9. identity (for multiplication)

10. associative property (for multiplication)

11. $4 + -17 + -3 + 29 = 4 + -20 + 29 = 4 + 9 = 13$

12. $-7 \cdot 5 \cdot -8 \cdot 12 = -7 \cdot -8 \cdot 5 \cdot 12 = 56 \cdot 60 = 3,360$

13. $8 \cdot -6 \cdot 3 \cdot -4 = 8 \cdot 3 \cdot -6 \cdot -4 = 24 \cdot 24 = 576$

14. $-14 + 8 + -31 + 27 + 6 + -4 = -8$

15. $-25(18 + -14) + 8(-9 + 12) = -25(4) + 8(3) = -100 + 24 = -76$

16. $7(-39 + 43) + 14 \cdot -2 = 7(4) + 14 \cdot -2 = 28 + -28 = 0$

17. $\dfrac{13}{5} + -\dfrac{17}{3} + -\dfrac{3}{5} + \dfrac{29}{3} = \dfrac{13}{5} + -\dfrac{3}{5} + -\dfrac{17}{3} + \dfrac{29}{3} = \dfrac{10}{5} + \dfrac{12}{3} = 2 + 4 = 6$

18. $4 \cdot (-0.6) \cdot (1.2) \cdot (-10)^2 = -288$

19. $-1.4 + 8.3 + -3.1 + 2.7 + -4.5 = 2$

20. $-14\left(-\dfrac{3}{8} + \dfrac{3}{7}\right) + \left(-\dfrac{7}{4} + \dfrac{7}{2}\right) = 7 - 6 = 1$

21. $\left(2^3\right) \cdot \left(2^2\right) = 2^5 = 32$

22. $\dfrac{5^4}{5^2} = 5^2 = 25$

23. $\left(3^2\right)^3 = 3^6 = 729$

24. $12^{-1} = \dfrac{1}{12}$

25. $\dfrac{4^5}{4^7} = 4^{-2} = \dfrac{1}{4^2} = \dfrac{1}{16}$

26. $6^{-3} \cdot 6^5 = 6^2 = 36$

27. $\dfrac{81^{14}}{81^{14}} = 81^0 = 1$

28. $\left(5^7\right)\left(5^{-8}\right) = 5^{-1} = \dfrac{1}{5}$

29. $\dfrac{9^4}{3 \cdot 9^3} = \dfrac{9}{3} = 3$

30. $\left(\dfrac{7^5 \cdot 11^{13}}{7^{12} \cdot 11^9}\right)^0 = 1$

31. $4 - 3 \cdot 4 + 3 = 4 - 12 + 3 = -8 + 3 = -5$

32. $(4-3) \cdot (4+3) = 1 \cdot 7 = 7$

33. $4 - 3 \cdot (4+3) = 4 - 3 \cdot 7 = 4 - 21 = -17$

34. $(4-3) \cdot 4 + 3 = 1 \cdot 4 + 3 = 4 + 3 = 7$

35. $5^2 - 2 \cdot (7-3) = 25 - 8 = 17$

36. $(66 - 54) \div 3 + 10 \div 5 - (6 - 2^2) = 4 + 2 - 2 = 4$

37. $5 - 2 + 2 \cdot (8-5) \div 2 = 3 + 3 = 6$

38. $17 - (8-3) \cdot (2+1) \div 5 = 14$

39. $8\left(2(4+3)^2 - 20 + 12\right) \div 4 = 720 \div 4 = 180$

40. $7 + 3^2 \div 2 - 6 = 5.5$

Chapter 2

1. $-4x$

2. $y + 18$

3. $\dfrac{z}{8}$

4. $9x - 3$

5. $19 + (-3)x = 19 - 3x$

6. $(a-b) \cdot -5 = -5(a-b)$

7. $19 - x$

8. $2x + 4 = 8 + y$

9. $3h + 1.50 = 12$

10. $n + (n + 15) = 42$

11. 2 factors: 3 and x

12. 3 factors: -6, y and y

13. 3 factors: 23, x and y

14. 6 factors: -48, x, x, x, y, and y

15. 3 terms

16. 2 terms

17. 1 term

18. 2 terms

19. 3 terms

20. 5 terms

21. $(x+3) + (9+x) = 2x + 12$

22. $(2x-1) + (3x+9) = 5x + 8$

23. $\left(x^2 + 5\right) + \left(x^2 + 6x\right) + (8x+3) = 2x^2 + 14x + 8$

24. $(2x+7) + (8x+4) + \left(-x^2 - 1\right) = -x^2 + 10x + 10$

25. $(2x-4) + (6x+1) + (3-x) = 7x$

26. $(y+5) + (2y-9) + (4y+10) + (y-6) = 8y$

27. $(8t-4)+(-t-1)+(13t-7)+(-2t-9)=18t-21$

28. $(x^2+4)+(-x^2-1)+(-3-5x)+(10x+9)=5x+9$

29. $(3x^2+2x+4)+(x^2-3x+10)=4x^2-x+14$

30. $(4x^2-5x-6)+(9+7x-4x^2)=2x+3$

31. $2(x-4)+7(x+4)=2x-8+7x+28=9x+20$

32. $-3(y+8)+8(y-3)=-3y-24+8y-24=5y-48$

33. $6(2t+5)+9(t-2)=12t+30+9t-18=21t+12$

34. $-7(3w-5)+4(w+5)=-21w+35+4w+20=-17w+55$

35. $x(x+4)-(x+4)=x^2+4x-x-4=x^2+3x-4$

36. $-8x(x-3)+6(x-4)=-8x^2+24x+6x-24=-8x^2+30x-24$

37. $5a(2a-7)-2(8-a)=10a^2-35a-16+2a=10a^2-33a-16$

38. $-4y(7-3y)-4(3-7y)=-28y+12y^2-12+28y=12y^2-12$

39. $3x(2x+3y)+3y(2x+3y)=6x^2+9xy+6xy+9y^2=6x^2+15xy+9y^2$

40. $2x^2(x^2-7x+9)+3x(x^2-7x+9)+2(x^2-7x+9)=2x^4-11x^3-x^2+13x+18$

41. $2x^2+x-28$

42. x^3-2x^2-x-6

43. $2x^3+15x^2+13x+3$

44. $6x^3-17x^2-12x+32$

45. $(x+3)\cdot(x+2)=x^2+2x+3x+6=x^2+5x+6$

46. $(x+4)\cdot(x-2)=x^2-2x+4x-8=x^2+2x-8$

47. $(2x+1)\cdot(x+7)=2x^2+14x+x+7=2x^2+15x+7$

48. $(4x-1)\cdot(2x+3)=8x^2+12x-2x-3=8x^2+10x-3$

49. $(5x-2)\cdot(2x-3)=10x^2-15x-4x+6=10x^2-19x+6$

50. $(x^2+x)\cdot(x-3)=x^3-3x^2+x^2-3x=x^3-2x^2-3x$

Chapter 3

1. Function. No rule.

2. Function. It's hard to see but this does follow a rule. Multiply the input by -3 and then add 4.

3. Function. Output = Input squared, minus 1.

4. Not a function. No rule.

5. Function. The outputs are the positive square roots of the inputs.

6. Domain: $\{-2,-1,0,1,2\}$, Range: $\{3,4,5,6,7\}$, Rule: Output = input + 5

7. Domain: $\{-2,-1,0,1,2\}$, Range: $\{4\}$, Rule: Output always equals 4, no matter what the input.

8. Domain: $\{0,2,4,6,8\}$, Range: $\{-3,1,5,9,13\}$, Rule: Output = twice the input minus 3.

9. Domain: $\{10,20,30,40,50\}$, Range: $\{1,11,21,31,41\}$, Rule: Output = input minus 9.

10. Domain: $\{-5,6,8,11,21\}$, Range: $\{-4,-1,-7,12,-5\}$, Rule: None

11. $f(3)=4-3\cdot3=4-9=-5$

12. $f(-5)=4-3(-5)=4+15=19$

13. $f(0)=4-3(0)=4-0=4$

14. $f(-1)=4-3(-1)=4+3=7$

15. $f(-100)=4-3(-100)=4+300=304$

16. If $g(x)=8x-2$ find
$g(-4)=8(-4)-2=-32-2=-34$

17. If $h(x)=x^2+5$ find $h(0)=0^2+5=0+5=5$

18. If $p(x)=x^2-3x+7$ find
$p(-1)=(-1)^2-3(-1)+7=1+3+7=11$

19. If $v(t)=25-8t$ find
$v(5)=25-8\cdot5=25-40=-15$

20. If $a(t)=9.8$ find $a(20)=9.8$ (Constant function)

21.

x	-2	-1	0	1	2
$f(x)$	-7	-4	-1	2	5

22.

x	-5	-3	-1	1	3	5
$g(x)$	-21	-5	3	3	-5	-21

23.

z	-4	-1	0	2	5
$p(z)$	1	7	9	13	19

24.

t	0	1	2	3	4	5
$v(t)$	-1	0	3	8	15	24

25.

t	-5	-4	-3	-2	-1	0
$a(t)$	15	12	9	6	3	0

26. $v(t)=5t-3$

27. $f(x)=-3x+7$

28. $g(x)=-x^2+5$

29. $a(t)=1.5t+2$

30. $p(x)=3x+10$

31.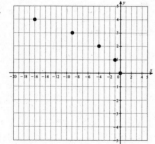

32.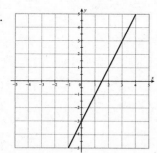

33.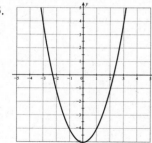

34.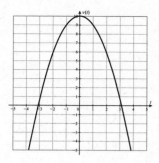

35.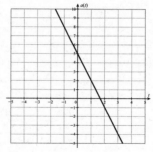

36.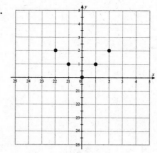

37.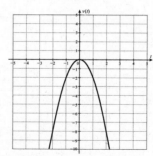

38.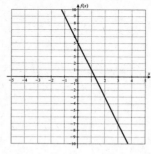

39.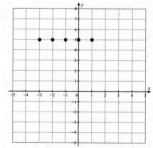

40.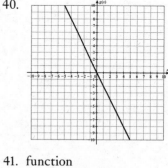

41. function

42. not a function

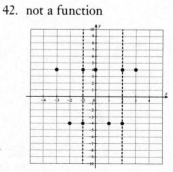

43. function

44. not a function

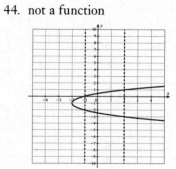

45. function

46. not a function

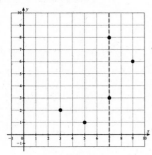

47. function

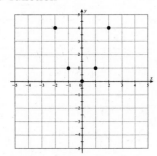

48. function

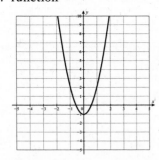

49. function

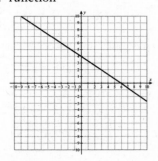

50. not a function

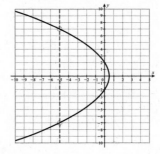

Chapter 4

1. $3x + 3 = 132$

2. $3y + 3 = y + 3$

3. $2t - 10 = 5 + t$

4. $2x - 8 = 2x - 1$

5. $-3a + 23 = 6a - 14$

6. $2x + 3 = 5x + 3$

7. $4v - 12 = -2v + 4$

8. $4x - 17 = 9x + 23$

9. $6t - 15 = 6t - 15$

10. $-3a - 19 = 4 - 3a$

11. $x = 7$

12. $t = 20$

13. $y = 27$

14. $48 = t$

15. $a = -25$

16. $w = 11$

17. $x = -9$

18. $t = 349$

19. $y = -120$

20. $z = 0$

21. $a = 7$

22. $z = -4$

23. $n = -7$

24. $y = 1$

25. $t = -5$

26. $x = 80$

27. $z = 2.5$

28. $w = -4$

29. $x = 3$

30. $y = -217$

31. $-4 = x$

32. $7 = t$

33. $y = 12$

34. $v = 45$

35. $7 = x$

36. no solution

37. $53 = x$

38. identity

39. $z = 0$

40. $-11.5 = x$

41. The number is 7.

42. The number is 4.

43. The two consecutive numbers are 26 and 27.

44. 5,000 storage containers were manufactured.

45. The garden measures 6 feet by 9 feet.

46. The price of a hot dog is $3.50.

47. Carlos and Dahlia will have the same amount in savings in 10 weeks.

48. The average price per e-book was $7.89.

49. Sophie bought 6 pounds of almonds.

50. The number is 8.

Chapter 5

1. $x < 12$

2. $r \le 4\%$ or $r \le 0.04$

3. $h > 48$

4. $p \ge 20$

5. $I \le 12,000$

Sample answers for problems 6 through 10.

6. A number, x, is at least 16.

7. A number, y, is less than 12.

8. A number, t, exceeds 22.

9. A number, z, is no more than 100.

10. A number, a, is more than 0.

11. $x > -3$

12. $y < 6$

13. $x \ge 4$

14. $t > -4.5$

15. $y \le 1.5$

16. $x = -4$ is part of the solution

17. $x = -6$ is not part of the solution

18. $y = 1$ is not part of the solution

19. $x = 5$ is not part of the solution

20. $t = 10$ is part of the solution

21. $x > 5$

22. $t \ge -8$

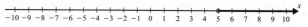

23. $y < -4$

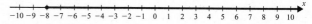

24. $z \ge 7$

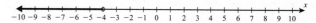

25. $a \le 9$

26. $x \le 4$

27. $-2 > x$

28. $x \le -4$

29. $y > -1$

30. $t \le 0$

31. $y > 3$ or $y < -2$

32. $1 \le z \le 3$

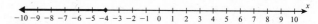

33. $18 \le x$ or $x < 7$

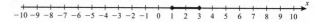
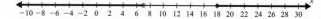

34. $3 < t < 4$

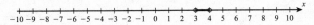

35. $17 < 4y + 5 \leq 1$ has no solution.

36. $-8 \leq t$ or $t > -5$ is equivalent to $t \geq -8$

37. $-4 \leq x \leq 6$

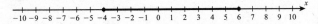

38. $5 < x$ or $x < 3.5$

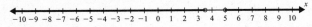

39. $-4 > t \geq -5$

40. $2 \leq y < 4$

Chapter 6

Answers for problems 1 through 5 are shown on the graph.

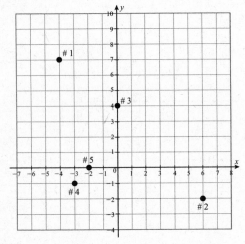

6. $(4, 1)$ or $(8, 9)$

7. $(0, 4)$ or $(0, -3)$

8. $(3, -1)$ or $(7, -5)$

9. $(-6, -3)$ or $(-8, -9)$

10. $(-5, 0)$ or $(2, 0)$

Sample tables are shown for problems 11 through 20.

11. $f(x) = 3x - 4$

x	-4	-3	-2	-1	0	1	2	3	4
y	-16	-13	-10	-7	-4	-1	2	5	8

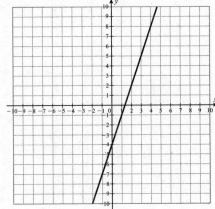

12. $g(x) = \frac{1}{2}x + 4$

x	-8	-6	-4	-2	0	2	4	6
y	0	1	2	3	4	5	6	7

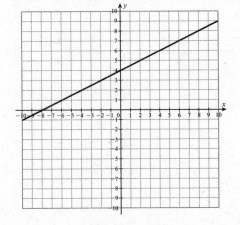

13. $y = 2x - 3$

x	-4	-3	-2	-1	0	1	2	3
y	-11	-9	-7	-5	-3	-1	1	3

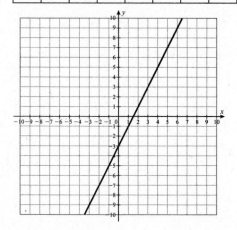

15. $y = -x + 5$

x	-4	-3	-2	-1	0	1	2	3
y	9	8	7	6	5	4	3	2

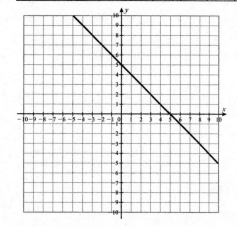

14. $g(x) = x + 5$

x	-4	-3	-2	-1	0	1	2	3
y	1	2	3	4	5	6	7	8

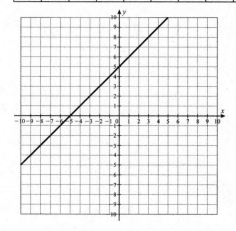

16. $f(x) = -2x + 7$

x	-4	-3	-2	-1	0	1	2	3
y	15	13	11	9	7	5	3	1

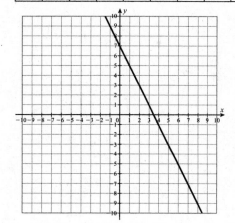

17. $g(x) = -\dfrac{2}{3}x + 6$

x	-12	-9	-6	-3	0	3	6	9
y	14	12	10	8	6	4	2	0

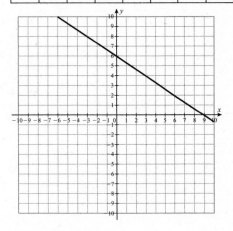

19. $f(x) = -2x$

x	-4	-3	-2	-1	0	1	2	3
y	8	6	4	2	0	-2	-4	-6

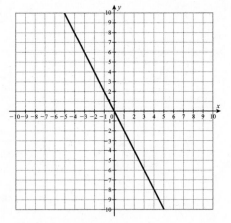

18. $y = 4 - 3x$

x	-4	-3	-2	-1	0	1	2	3
y	16	13	10	7	4	1	-2	-5

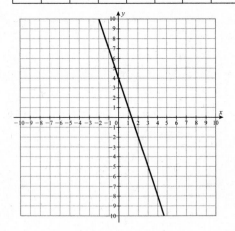

20. $g(x) = -\dfrac{3}{2}x + 5$

x	-8	-6	-4	-2	0	2	4	6
y	11	17	11	8	5	2	-1	-4

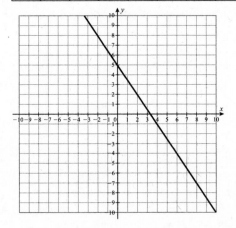

21. x-intercept: (3, 0), y-intercept: (0, -6)

22. x-intercept: (2, 0), y-intercept: (0, 6)

23. x-intercept: (-1, 0), y-intercept: (0, 4)

24. x-intercept: (5, 0), y-intercept: (0, 5)

25. x-intercept: (-4, 0), y-intercept: (0, 4)

26. *x*-intercept: (4, 0), *y*-intercept: (0, 8)

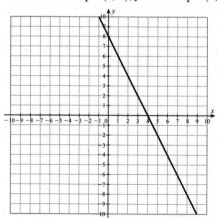

27. *x*-intercept: (2, 0), *y*-intercept: (0, -6)

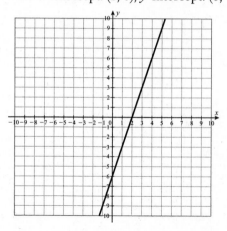

28. *x*-intercept: (6, 0), *y*-intercept: (0, -6)

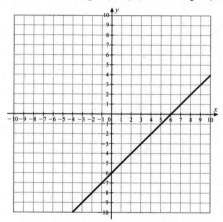

29. *x*-intercept: (-5, 0), *y*-intercept: (0, 5)

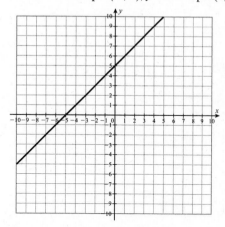

30. *x*-intercept: (8, 0), *y*-intercept: (0, -4)

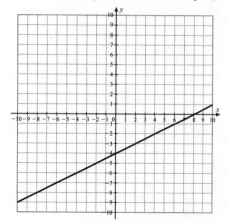

31. Slope: $\frac{2}{1}$, *y*-intercept: (0, -5)

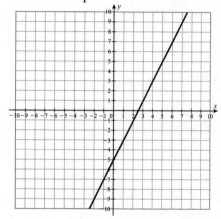

32. Slope: $\dfrac{3}{1}$, y-intercept: (0, -7)

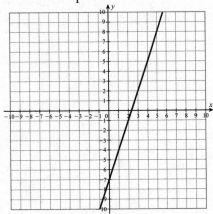

35. Slope: $\dfrac{2}{3}$, y-intercept: (0, -4)

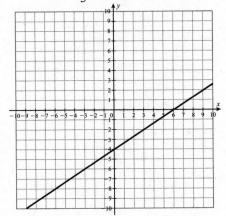

33. Slope: $\dfrac{-4}{1}$, y-intercept: (0, 9)

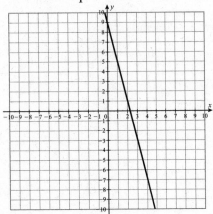

36. Slope: $-\dfrac{4}{3}$, y-intercept: (0, 7)

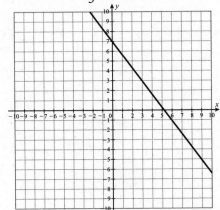

34. Slope: $\dfrac{1}{2}$, y-intercept: (0, -3)

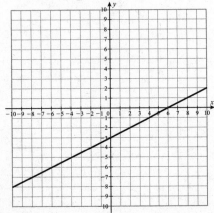

37. Slope: $\dfrac{-1}{1}$, y-intercept: (0, 5)

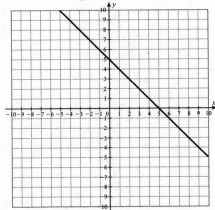

38. Slope: $\frac{2}{1}$, y-intercept: (0, -9)

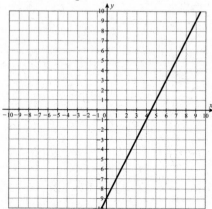

39. Slope: $\frac{2}{1}$, y-intercept: (0, 3)

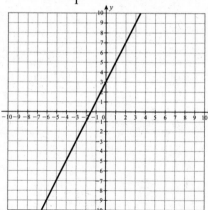

40. Slope: $\frac{-2}{1}$, y-intercept: (0, 1)

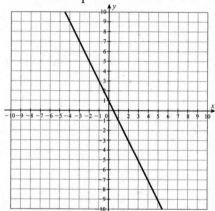

41. direct variation

42. No, direct variation doesn't have constant term.

43. No, $xy = 12$ is equivalent to $y = \frac{12}{x}$. Direct variation cannot have a variable in the denominator.

44. direct variation

45. No, the graph of a direct variation equation passes through the origin.

46. $k = 45$

47. $k = 2\pi$

48. The equation is $F = 50a$.

49. Two pounds of potatoes cost $2.98.

50. At 30 degrees Celsius, the volume will be 1,800 cubic centimeters.

Chapter 7

1. $y = \frac{-3}{2}x + 3$

2. $y = 2x + \frac{1}{2}$

3. $y = -\frac{1}{5}x + 2$

4. $y = 3x - 9$

5. $y = -4x + 2$

6. $y = \frac{1}{2}x - 2$

7. $y = -6x + 3$

8. $y = \frac{3}{2}x - \frac{5}{2}$

9. $y = -\frac{9}{7}x - 9$

10. $y = 10x + 10$

11. $6x - y = 7$

12. $3x - y = -13$

13. $3x + y = 5$

14. $4x + y = 11$

15. $8x - 4y = 32$

16. $x - 2y = 2$

17. $4x + 5y = 3$

18. $6x + 2y = -19$

19. $3x - 2y = 7$

20. $3x + 5y = 28$

21. $y - 7 = -2(x + 5)$ point-slope form

22. $3x - 4y = 12$ standard form

23. $y = 2x - 6$ slope-intercept form

24. $x + y = 10$ standard form

25. $y + 2 = \frac{3}{5}(x - 4)$ point-slope form

26. $y = -3x + 0$ or simply $y = -3x$

27. $y = 5x - 9$

28. $y = \frac{1}{2}x + 5$

29. $y = -\frac{4}{3}x - 6$

30. $y = 1x + 3$ or simply $y = x + 3$

31. $y = -4x + 3$

32. $y = 5x + 16$

33. $y = \frac{1}{3}x + 6$

34. $y = -\frac{5}{4}x - 5$

35. $y = -x + 5$

36. $y = 2x - 3$

37. $y = \frac{3}{2}x + 15$

38. $y = x + 1$

39. $y = \frac{2}{5}x + 3$

40. $y = 3x - 23$

41. $m = \dfrac{-7 - 3}{8 - 6} = \dfrac{-10}{2} = -5$

42. $m = \dfrac{9 - 5}{1 + 3} = \dfrac{4}{4} = 1$

43. $m = \dfrac{-2 + 5}{-1 + 4} = \dfrac{3}{3} = 1$

44. $m = \dfrac{8 + 4}{-4 - 2} = \dfrac{12}{-6} = -2$

45. $m = \dfrac{7 - 0}{0 - 7} = \dfrac{7}{-7} = -1$

46. $m = \dfrac{20 + 10}{-5 - 5} = \dfrac{30}{-10} = -3$, $y + 10 = -3(x - 5)$
    or $y - 20 = -3(x + 5)$

47. $m = \dfrac{-4 + 9}{6 + 4} = \dfrac{5}{10} = \dfrac{1}{2}$, $y + 9 = \dfrac{1}{2}(x + 4)$ or
    $y + 4 = \dfrac{1}{2}(x - 6)$

48. $m = \dfrac{7 + 5}{1 - 4} = \dfrac{12}{-3} = -4$, $y + 5 = -4(x - 4)$ or
    $y - 7 = -4(x - 1)$

49. $m = \dfrac{3 - 5}{-3 - 3} = \dfrac{-2}{-6} = \dfrac{1}{3}$, $y - 5 = \dfrac{1}{3}(x - 3)$ or
    $y - 3 = \dfrac{1}{3}(x + 3)$

50. $m = \dfrac{\frac{33}{4} - \frac{21}{2}}{1 + 2} = \dfrac{\frac{33}{4} - \frac{42}{4}}{3} = \dfrac{-\frac{9}{4}}{3} \cdot \dfrac{4}{4} = \dfrac{-9}{12} = \dfrac{-3}{4}$,
    $y - \dfrac{21}{2} = -\dfrac{3}{4}(x + 2)$ or $y - \dfrac{33}{4} = -\dfrac{3}{4}(x - 1)$

51. Use $m = 2$. $y + 3 = 2(x - 6)$ or $y = 2x - 15$.

52. $3y = x + 12 \Rightarrow y = \dfrac{1}{3}x + 4$. Use $m = \dfrac{1}{3}$.
    $y - 2 = \dfrac{1}{3}(x + 3)$ or $y = \dfrac{1}{3}x + 3$.

53. Use $m = 4$. $y + 3 = 4(x + 4)$ or $y = 4x + 13$.

54. Use $m = -6$. $y - \dfrac{1}{2} = -6\left(x - \dfrac{1}{2}\right)$ or
    $y = -6x + 3\dfrac{1}{2}$.

55. Use $m = \dfrac{4}{3}$. $y - 4 = \dfrac{4}{3}\left(x - \dfrac{3}{2}\right)$ or $y = \dfrac{4}{3}x + 2$.

56. Use $m = -\frac{1}{2}$. $y - 7 = -\frac{1}{2}(x - 4)$ or $y = -\frac{1}{2}x + 9$.

57. Use $m = \frac{1}{3}$. $y + 2 = \frac{1}{3}(x - 6)$ or $y = \frac{1}{3}x - 4$.

58. $x - 2y = -12 \Rightarrow y = \frac{1}{2}x + 6$. Use $m = -2$. $y - 7 = -2(x + 1)$ or $y = -2x + 5$.

59. Use $m = -3$. $y - \frac{2}{3} = -3\left(x + \frac{1}{3}\right)$ or $y = -3x - \frac{1}{3}$.

60. Use $m = -1$. $y + 3 = -1(x - 4)$ or $y = -x + 1$.

Chapter 8

1. $|9 - n| = 6$ or $|n - 9| = 6$

2. $|t - (-4)| = 10$ or $|-4 - t| = 10$

3. $|x - 5| = 12$ or $|5 - x| = 12$

4. $2|y - (-4)| = 12$ or $2|-4 - y| = 12$

5. $|z - 12| = z + 4$ or $|z - 12| = z + 4$

Sample sentences are given for problems 6 through 10.

6. The number 7 is 3 units from x.

7. The distance between t and -2 is 9 units.

8. The number y is 11 units from 8.

9. Five times the distance between x and 4 is 20.

10. The distance between z and 7 is 3 less than z.

11. $x = 3$ or $x = \frac{5}{3}$

12. $t = -1$ or $t = -4$

13. $y = 9$ or $y = 7$

14. $x = 8$ or $x = 18$

15. $|11 + 8x| = -17$ has no solution. Absolute value is never negative.

16. $x = \frac{8}{3}$ or $x = -2$

17. $y = 3$ or $y = 1$

18. $x = 2$ or $x = -4$

19. $x = \frac{2}{3}$ or $x = 2$

20. $\frac{5}{3} = x$ or $1 = x$

21. $\frac{1}{3} > x$ or $x > 2$

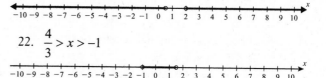

22. $\frac{4}{3} > x > -1$

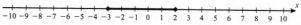

23. $-2.8 \le y \le 2$

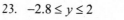

24. $\frac{8}{7} \ge z$ or $z \ge 2$

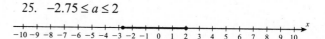

25. $-2.75 \le a \le 2$

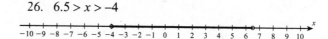

26. $6.5 > x > -4$

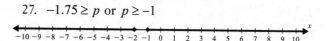

27. $-1.75 \ge p$ or $p \ge -1$

28. $|9t + 5| \le -2$ has no solution. Absolute value is never negative.

29. $-21 < x < 11$

30. $|7y - 23| \ge 0$ is true for all values of the variable. The solution set is all real numbers.

31. $x = 36$ or $x = -8$

32. $y = -41$ or $y = 7$

33. The number n is between 6 and 12, inclusive.

34. Paul's salary is between $86,500 and $78,500.

35. The points are (0, -10) or (0, 2).

36. The possible points are (-25, 0) or (1, 0).

37. The points are (0, -7) or (0, 13).

38. $96.6 > t$ or $t > 100.6$

39. The range of acceptable voltages is $110 < V < 120$.

40. The candidate should receive between 42% and 50% of the vote.

43. $y = |x + 1|$

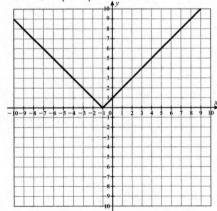

41. $y = |x - 3|$

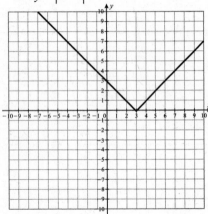

44. $y = |x| - 5$

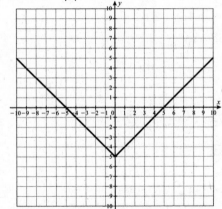

42. $y = |x| + 1$

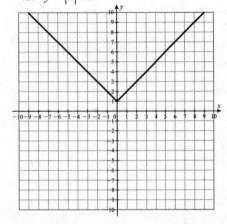

45. $y = |2x|$

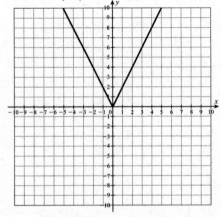

46. $y = -3|x|$

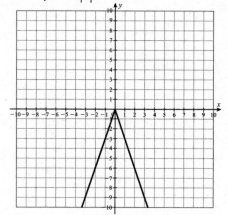

47. $y = |2x - 5|$

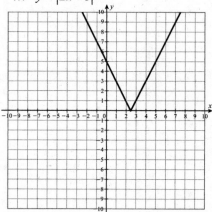

48. $y = |7 - 4x|$

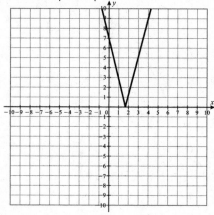

49. $y = |3x - 8|$

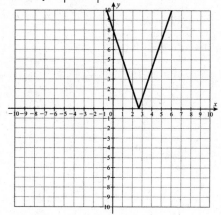

50. $y = |12 - 5x|$

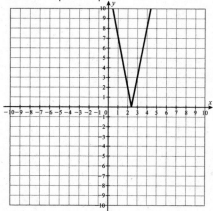

Chapter 9

1. $y = 12$ Horizontal line

2. $x = 5y$ Oblique line

3. $y = 12 + x$ Oblique line

4. $x = 7$ Vertical line

5. $x - 6 = 0$ Vertical line

6. $x = -4$

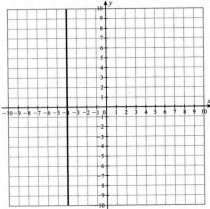

9. $y + 7 = 4$

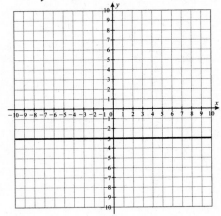

7. $y = -3$

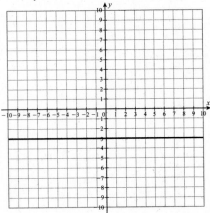

10. $x = 1$

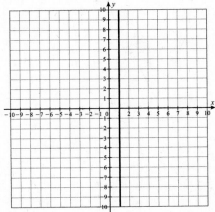

8. $x + 4 = 6$

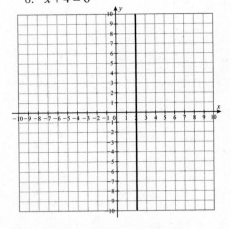

11. $y \leq 2x - 1$

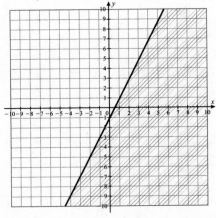

12. $y > \dfrac{1}{2}x + 3$

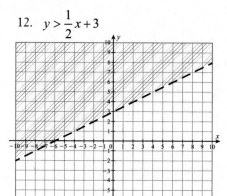

15. $x < 3$

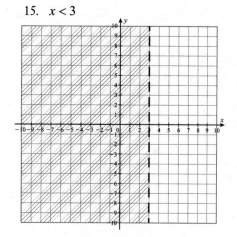

13. $y < -3x + 5$

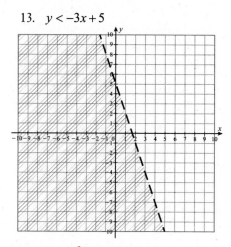

16. $y - 5 \geq 0$

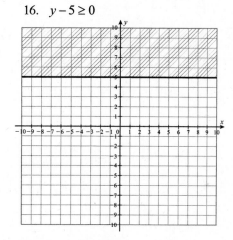

14. $y \geq -\dfrac{2}{3}x + 6$

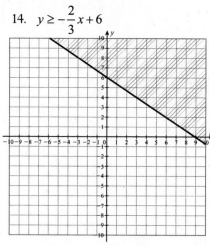

17. $x + y \leq 8$

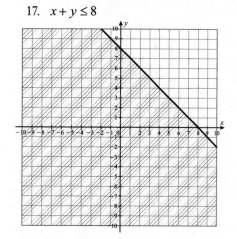

18. $2x - y > -6$

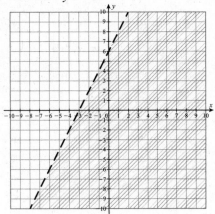

21. $y < |x - 7|$

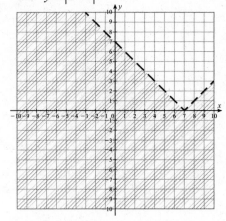

19. $4x - 3y < 12$

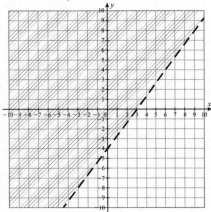

22. $y \geq |x + 3|$

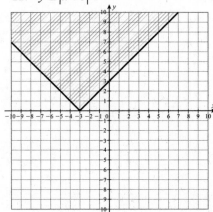

20. $x \leq 2y + 8$

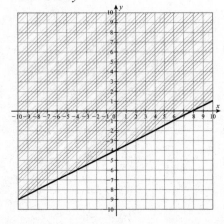

23. $y > |x| - 4$

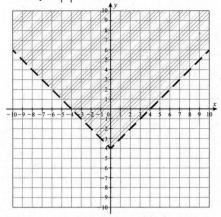

24. $y \le |x| + 1$

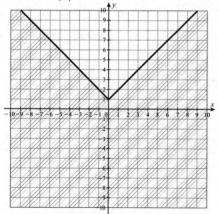

27. $y < \left|\dfrac{1}{2}x - 3\right|$

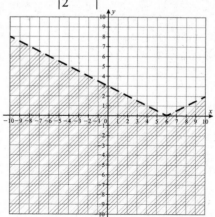

25. $y > \left|\dfrac{3}{2}x\right|$

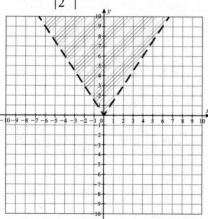

28. $y \le \left|\dfrac{1}{2}x\right| - 3$

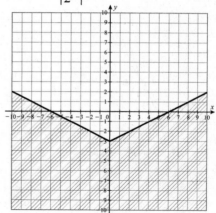

26. $y \ge |2x| - 3$

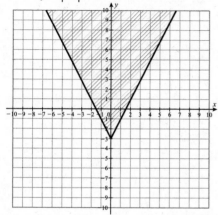

29. $y \ge -2|x| + 5$

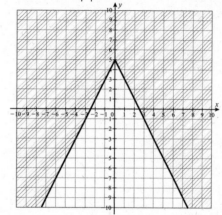

30. $y > -3|x+4| + 5$

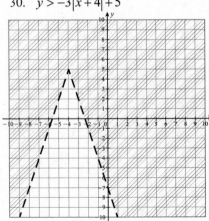

Chapter 10

1. $y = 2x + 3$

$y = 3x - 1$

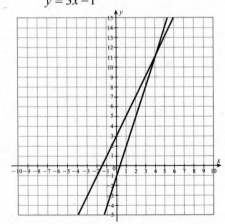

2. $y = -\dfrac{1}{2}x + 4$

$y = \dfrac{3}{4}x - 1$

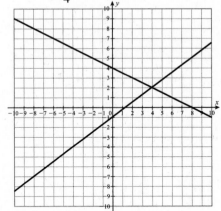

3. $x + y = 9$

$x - y = 1$

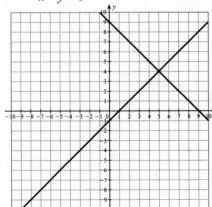

4. $3x - y = 5$
   $6x - 2y = 8$

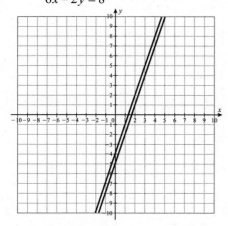

5. $2x - 5y = -44$
   $y = 6 - x$

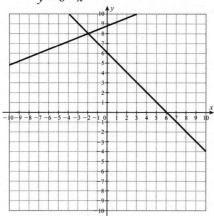

6. $y = 3x - 7$
   $2x - 3y = 14$

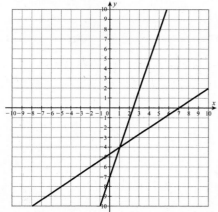

7. $y = 3x - 5$
   $y = -\dfrac{1}{3}x + 5$

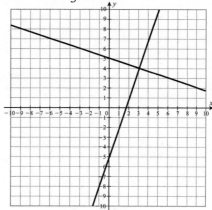

8. $2x + y = -8$
   $3x - 5y = 14$

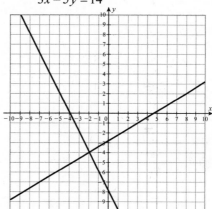

9. $x + y = 6$
   $y = -x - 4$

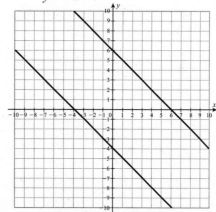

10. $3x - 2y = 5$
     $x + 4y = -3$

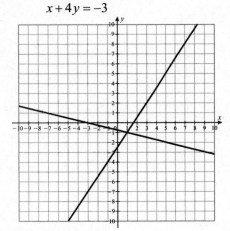

11. $x = 1, y = 10$

12. $x = \dfrac{5}{2}, y = -2$

13. $x = 3, y = -3$

14. $x = -2, y = -5$

15. $a = 1, b = 6$

16. $x = 0, y = 9$

17. $x = 5, y = 2$

18. $x = 7, y = 3$

19. $x = -1, y = 2$

20. $a = -1, b = 2$

21. $x = 10, y = 7$

22. $x = 7, y = 1$

23. $x = 2, y = 3$

24. $x = 3, y = 0$

25. $x = 3, y = -3$

26. $x = 4, y = -3$

27. $x = 7, y = 3$

28. $x = 5, y = 2$

29. $x = 3, y = 4$

30. $x = -1, y = 2$

31. There are 7 quarters and 13 dimes.

32. There were 18 of the 14 cent stamps and 12 of the 17 cent stamps.

33. Use 20 pounds of $6 coffee and 40 pounds of $5.10 coffee.

34. Invest $2,600 at 4% and $3,400 at 7%.

35. Use 16 ml of 5% solution and 32 ml of 20% solution.

36. She traveled 6.4 hours at 50 mph and 1.6 hours at 20 mph.

37. Adult admission was $10 and children's admission was $3.50.

38. There were 7 nickels and 13 dimes.

39. Pens cost 80 cents and pencils cost 30 cents.

40. There were 110 adults and 140 children.

41. $y \le 2x + 1$
     $y \ge 4 - x$

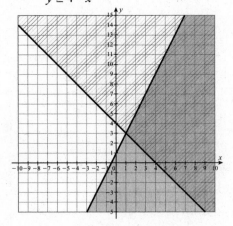

42. $y < 7 - 3x$
        $2x + y > 8$

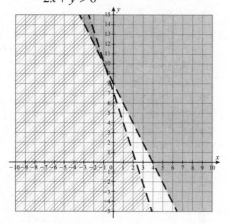

43. $3x - y < 1$
        $y \geq 2 - 3x$

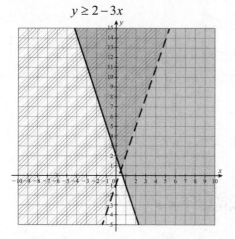

44. $5x + 2y \geq 12$
        $3x + 7 < y$

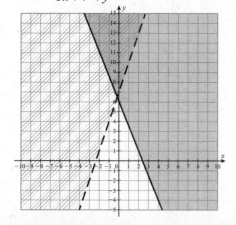

45. $3x + 4 \geq 4y$
        $x + 2y < 12$

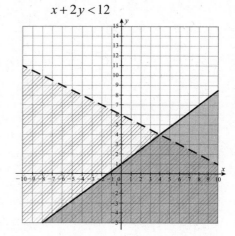

46. $x + y \geq 40$
        $x - y \leq 8$

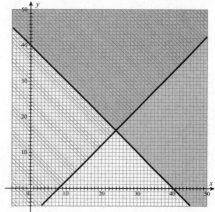

47. $x - y \leq 12$
        $x + y \geq 20$

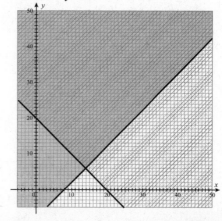

48. $x + y \leq 80$

$100x + 300y \geq 7500$

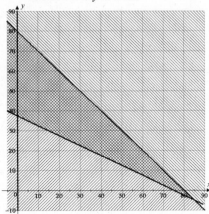

49. $15x + 20y \leq 400$

$12x + 30y \leq 300$

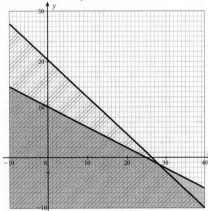

50. $2x + 9y \leq 25$

$3x + 4y \leq 36$

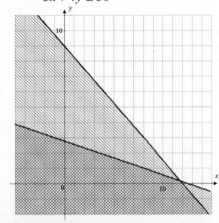

Chapter 11

1. $x^2 + 4x - 7 + \dfrac{3}{x}$ is not a polynomial.

Division by a variable is not permitted, because they would be represented by negative exponents.

2. $3y^7 - 4y^3 + 8$ is a polynomial.

3. $2t^3 + 4t^2 - 8t + 4\sqrt{t}$ is not a polynomial because there is a variable under a radical. Radicals are the equivalent of fractional exponents.

4. $9z - 6z^2 + 2 = -6z^2 + 9z + 2$

5. $4x^2 - 3x + x^3 - 2 = x^3 + 4x^2 - 3x - 2$

6. $8t - 2 + t^2 = t^2 + 8t - 2$

7. $9y - 3$ is a binomial of degree 1.

8. $x^2 + 5x - 9$ is a trinomial of degree 2.

9. $-\dfrac{1}{2}z^3$ is a monomial of degree 3.

10. $12a^9 + 8a^{12} - 4a^7 + 6a^4 + 2$ is a polynomial of degree 12.

11. $\left(5x^3\right) + \left(7x^2\right) + \left(2x^3\right) = 7x^3 + 7x^2$

12. $\left(4x - 7\right) + \left(9 - 2x\right) = 2x + 2$

13. $\left(3b + b^2\right) + \left(5b - 6b^2\right) = 8b - 5b^2$

14. $\left(3x^2 - 2x - 5\right) + \left(-5x^2 - 4x - 6\right) = -2x^2 - 6x - 11$

15. $\left(x^2 - 3x + 4\right) + \left(-7x^2 - 5x\right) = -6x^2 - 8x + 4$

16. $\left(x^2 - 8\right) + \left(5x^2 + 6x - 1\right) = 6x^2 + 6x - 9$

17. $\left(2y^3 + 3y^2 - y\right) + \left(5y^3 - 6y^2 + 3y\right) + \left(2y^3 - y^2 - 7y\right) = 9y^3 - 4y^2 - 5y$

18. $\left(2y^2 + 6y - 5\right) + \left(7y - 3y^2 + 6\right) + \left(8 - 9y + 2y^2\right) = y^2 + 4y + 9$

19. $\left(3t^2 - 2t + 6\right) + \left(7t - t^2 - 1\right) + \left(4 - 6t^2 + t\right) = -4t^2 + 6t + 9$

20. $\left(2x + 3x^2\right) + \left(-5x^3 - 4x\right) + \left(7x^3 - x^2\right) = 2x^3 + 2x^2 - 2x$

21. $\left(8x - 3\right) - \left(4 - 7x\right) = 15x - 7$

22. $\left(9x^2 - 7x + 5\right) - \left(-2x^2 + 6x + 3\right) = 11x^2 - 13x + 2$

23. $\left(2x^2 - 3x\right) - \left(4x^2 + 5x - 2\right) = -2x^2 - 8x + 2$

24. $\left(2y^2 + 7y - 4\right) - \left(3y^2 - 5y + 2\right) = -y^2 + 12y - 6$

25. $\left(5t^2 - 7t + 2\right) - \left(3t^2 - 9t - 1\right) = 2t^2 + 2t + 3$

26. $\left(2a^2 + 6 - 3a\right) - \left(8 + 9a - a^2\right) = 3a^2 - 12a - 2$

27. $\left(5x^2 - 2x + 3\right) - \left(2x^2 - x + 5\right) = 3x^2 - x - 2$

28. $\left(y^3 + 5\right) - \left(y^2 - 3\right) = y^3 - y^2 + 8$

29. $\left(4x^3 + 5x^2 - 7\right) - \left(2x^2 - 8x - 1\right) = 4x^3 + 3x^2 + 8x - 6$

30. $\left(2x^2 + 3x\right) - \left(5x^2 + 2\right) = -3x^2 + 3x - 2$

31. $\left(8x - 3\right) + \left(4 - 7x\right) - \left(4x - 7\right) + \left(9 - 2x\right) = -5x + 17$

32. $\left(y^3 + 5\right) + \left(y^2 - 3\right) + \left(3y + y^2\right) - \left(5y - 6y^2\right) = y^3 + 8y^2 - 2y + 2$

33. $\left(2x^2 - 3x\right) - \left(3x^2 - 2x - 5\right) - \left(4x^2 + 5x - 2\right) + \left(-5x^2 - 4x - 6\right) = -10x^2 - 10x + 1$

34. $\left(x^2 - 3x + 4\right) - \left(-7x^2 - 5x\right) - \left(2x^2 - 3x\right) + \left(4x^2 + 5x - 2\right) = 10x^2 + 10x + 2$

35. $\left(3x^2 + 2x - 1\right) + \left(5x^2 - 7x + 6\right) - \left(7x^2 - 15x - 3\right) = x^2 + 10x + 8$

36. $\left(2x^2 + 3x\right) + \left(5x^2 + 6x - 1\right) - \left(5x^2 + 2\right) - \left(x^2 - 8\right) = x^2 + 9x + 5$

37. $\left(8a^3 - 4a^2 + 6a + 2\right) - \left(2a^2 + 6 - 3a\right) + \left(8 + 9a - a^2\right) = 8a^3 - 7a^2 + 18a + 4$

38. $\left(4x^3 + 5x^2 - 7\right) - \left(7x^3 - x^2\right) - \left(2x^2 - 8x - 1\right) + \left(2x + 3x^2\right) + \left(-5x^3 - 4x\right) = -8x^3 + 7x^2 + 6x - 6$

39. $\left(3t^2 - 9t - 1\right) - \left(3t^2 - 2t + 6\right) + \left(4 - 6t^2 + t\right) = -6t^2 - 6t - 3$

40. $\left(2x^2 + 3x\right) - \left(-5x^2 - 4x - 6\right) + \left(x^2 - 2x - 5\right) + \left(4x^2 - 5x - 2\right) = 12x^2 - 1$

Chapter 12

1. $5a\left(2a^2 + 3a\right) = 10a^3 + 15a^2$

2. $-3b^2\left(2b^2 - 3b + 5\right) = -6b^4 + 9b^3 - 15b^2$

3. $2x^2\left(11x^2 - 3x - 7\right) = 22x^4 - 6x^3 - 14x^2$

4. $-3x^3\left(2 + 4x - 5x^2\right) = -6x^3 - 12x^4 + 15x^5$

5. $3y^4\left(2y^6 - y^2\right) = 6y^{10} - 3y^6$

6. $8x\left(x^2 - 3\right) = 8x^3 - 24x$

7. $-5a^3\left(2a^4 - 4a^2 + 6\right) = -10a^7 + 20a^5 - 30a^3$

8. $7x^5\left(9 - 3x^3\right) = 63x^5 - 21x^8$

9. $-6x^4\left(x^2 + x + 3\right) = -6x^6 - 6x^5 - 18x^4$

10. $-x^3\left(x^5 - x^4 + x^3 - x^2 + x - 1\right) = -x^8 + x^7 - x^6 + x^5 - x^4 + x^3$

11. $\left(y + 6\right)\left(y - 1\right) = y^2 + 5y - 6$

12. $(x+5)(x+2)=x^2+7x+10$

13. $(t-3)(t-2)=t^2-5t+6$

14. $(b-4)(b+6)=b^2+2b-24$

15. $(y+3)(3y-1)=3y^2+8y-3$

16. $(3x-8)(4x+5)=12x^2-17x-40$

17. $(3x+1)(2x-3)=6x^2-7x-3$

18. $(2x-7)(4x+9)=8x^2-10x-63$

19. $(10a-3)(8a-5)=80a^2-74a+15$

20. $(x^2+9)(x^2-4)=x^4+5x^2-36$

21. $(t-4)^2=t^2-8t+16$

22. $(y-7)(y+7)=y^2-49$

23. $(2x-1)^2=4x^2-4x+1$

24. $\left(x+\dfrac{1}{2}\right)\left(x-\dfrac{1}{2}\right)=x^2-\dfrac{1}{4}$

25. $(2t-5)(2t+5)=4t^2-25$

26. $(3x+5)^2=9x^2+30x+25$

27. $(5a-3)^2=25a^2-30a+9$

28. $(x^2-5)^2=x^4-10x^2+25$

29. $(x^2+1)(x^2-1)=x^4-1$

30. $(5t^2-2t)(5t^2+2t)=25t^4-4t^2$

31. $(4x-3)(3x^2+x-6)=12x^3-5x^2-27x+18$

32. $(x+5)(x^2+2x+3)=x^3+7x^2+13x+15$

33. $(2y-3)(y^2-3y+2)=2y^3-9y^2+13y-6$

34. $(2a-5)(a^3-5a^2+a-3)=2a^4-15a^3+27a^2-11a+15$

35. $(3x-2)(x^2-3x-6)=3x^3-11x^2-12x+12$

36. $(2y-3)(4y^3-y^2+2y+5)=8y^4-14y^3+7y^2+4y-15$

37. $(x-2)(x^3+5x-7)=x^4-2x^3+5x^2-17x+14$

38. $(3a+1)(a^2-a)=3a^3-2a^2-a$

39. $(2x^3-1)(6x^4+3x^2+1)=12x^7+6x^5-6x^4+2x^3-3x^2-1$

40. $(t-4)(t^2+4t+16)=t^3-64$

41. $(8x+24)\div8=x+3$

42. $(28x^2+84x)\div7x=4x+12$

43. $(6x^5-12x^4+9x^3)\div3x=2x^4-4x^3+3x^2$

44. $(15y^3-20y^2-10y)\div-5y=-3y^2+4y+2$

45. $(8x^4-12x^3)\div(4x^2)=2x^2-3x$

46. $(15t^6-21t^3)\div(-3t^2)=-5t^4+7t$

47. $(16y^5-24y^3)\div(-8y^3)=-2y^2+3$

48. $(14x^6-42x^4+56x^2)\div(-14x^2)=-x^4+3x^2-4$

49. $(-9a^4+27a^3-81a^2)\div(-9a^2)=a^2-3a+9$

50. $(24x^7-16x^5-48x^4+36x^3)\div(-4x^2)=-6x^5+4x^3+12x^2-9x$

51. $(x^2+11x+30)\div(x+5)=x+6$

52. $(x^2+10x+24)\div(x+4)=x+6$

53. $(z^2-15z+56)\div(z-8)=z-7$

54. $(b^2-3b-28)\div(b+4)=b-7$

55. $(6x^2+13x+6)\div(3x+2)=2x+3$

56. $(7y^2-3y-4)\div(y-1)=7y+4$

57. $(9x^2-42x+45)\div(3x-8)=3x-6-\dfrac{3}{3x-8}$

58. $(2b^2-7b+3)\div(b-3)=2b-1$

59. $(x^4-2x^2+1)\div(x^2-2x+1)=x^2+2x+1$

60. $(3x^4-17x^2+10)\div(x^2-5x)=3x^2+15x+58+\dfrac{290x+10}{x^2-5x}$

Chapter 13

1. $12x^2 - 18x = 6x(2x - 3)$

2. $21t^2 - 35 = 7(3t - 5)$

3. $16y^3 - 56y^2 + 72y = 8y(2y^2 - 7y + 9)$

4. $5ax^2 - 25ax + 15a = 5a(x^2 - 5x + 3)$

5. $32x^5 - 48x^4 + 16x^3 = 16x^3(2x^2 - 3x + 1)$

6. $11 - 77t + 22t^2 = 11(1 - 7t + 2t^2)$

7. $\pi r^2 + \pi rh = \pi r(r + h)$

8. $15x^2 - 3x - 5$ is prime.

9. $y^2 + 2xy = y(y + 2x)$

10. $4a^3 + 8a^2 - 24a = 4a(a^2 + 2a - 6)$

11. $x^2 + 7x + 12 = (x + 3)(x + 4)$

12. $y^2 - 6x + 8 = (y - 4)(y - 2)$

13. $t^2 + 9t + 20 = (t + 4)(t + 5)$

14. $y^2 - 2y + 1 = (y - 1)(y - 1)$

15. $x^2 + 12x + 27 = (x + 3)(x + 9)$

16. $t^2 - t - 12 = (t - 4)(t + 3)$

17. $x^2 + 2x - 15 = (x + 5)(x - 3)$

18. $y^2 - 8y - 20 = (y - 10)(y + 2)$

19. $t^2 + 10t - 11 = (t + 11)(t - 1)$

20. $x^2 - 12x - 28 = (x + 2)(x - 14)$

21. $8x^2 + 2x - 1 = (4x - 1)(2x + 1)$

22. $3a^2 + 2a - 8 = (3a - 4)(a + 2)$

23. $7y^2 + 20y - 3 = (7y - 1)(y + 3)$

24. $6x^2 + 5x - 6 = (2x + 3)(3x - 2)$

25. $3t^2 - 11t - 15$ is prime.

26. $2x^2 - 15x + 7 = (2x - 1)(x - 7)$

27. $3y^2 - 14y - 5 = (3y + 1)(y - 5)$

28. $10t^2 - 23t + 12 = (2t - 3)(5t - 4)$

29. $4x^2 - 17x - 15 = (x - 5)(4x + 3)$

30. $10x^2 + 7x - 12 = (2x + 3)(5x - 4)$

31. $x^2 - 64 = (x + 8)(x - 8)$

32. $y^2 + 12y + 36 = (y + 6)^2$

33. $4x^2 - 121 = (2x + 11)(2x - 11)$

34. $a^2 - 16a + 64 = (a - 8)^2$

35. $9y^2 - 49 = (3y + 7)(3y - 7)$

36. $25t^2 + 30t + 9 = (5t + 3)^2$

37. $81t^2 - 16 = (9t + 4)(9t - 4)$

38. $16x^2 - 56x + 49 = (4x - 7)^2$

39. $100x^2 - 1 = (10x + 1)(10x - 1)$

40. $36z^2 + 60z + 25 = (6z + 5)^2$

Chapter 14

1. $\sqrt{8} = 2\sqrt{2}$

2. $\sqrt{27} = 3\sqrt{3}$

3. $\sqrt{128} = 8\sqrt{2}$

4. $\sqrt{5x^2} = |x|\sqrt{5}$

5. $\sqrt{50x^2} = 5|x|\sqrt{2}$

6. $5\sqrt{32x^3} = 20|x|\sqrt{2x}$

7. $\sqrt{98t^4} = 7t^2\sqrt{2}$

8. $\sqrt{x^3y^2} = |xy|\sqrt{x}$

9. $9\sqrt{4a^3b^5} = 18|a|b^2\sqrt{ab}$

10. $5x\sqrt{12x^3} = 10x|x|\sqrt{3x}$

11. $\dfrac{8}{\sqrt{2x}} \cdot \dfrac{\sqrt{2x}}{\sqrt{2x}} = \dfrac{8\sqrt{2x}}{2x} = \dfrac{4\sqrt{2x}}{x}$

12. $\dfrac{9}{2\sqrt{12}} \cdot \dfrac{\sqrt{12}}{\sqrt{12}} = \dfrac{9\sqrt{12}}{2 \cdot 12} = \dfrac{9\sqrt{12}}{24} = \dfrac{3\sqrt{3}}{4}$

13. $\dfrac{27}{4\sqrt{18}} \cdot \dfrac{\sqrt{18}}{\sqrt{18}} = \dfrac{27\sqrt{18}}{4 \cdot 18} = \dfrac{3\sqrt{18}}{4 \cdot 2} = \dfrac{9\sqrt{2}}{8}$

14. $\dfrac{5\sqrt{2}}{4\sqrt{10}} \cdot \dfrac{\sqrt{10}}{\sqrt{10}} = \dfrac{5\sqrt{20}}{4 \cdot 10} = \dfrac{\sqrt{5}}{4}$

15. $\dfrac{10}{\sqrt{50}} \cdot \dfrac{\sqrt{50}}{\sqrt{50}} = \dfrac{10\sqrt{50}}{50} = \sqrt{2}$

16. $\dfrac{1}{\left(3-\sqrt{3}\right)} \cdot \dfrac{\left(3+\sqrt{3}\right)}{\left(3+\sqrt{3}\right)} = \dfrac{3+\sqrt{3}}{6}$

17. $\dfrac{5}{\left(2+\sqrt{3}\right)} \cdot \dfrac{\left(2-\sqrt{3}\right)}{\left(2-\sqrt{3}\right)} = 5\left(2-\sqrt{3}\right)$

18. $\dfrac{1}{\left(\sqrt{3}-8\right)} \cdot \dfrac{\left(\sqrt{3}+8\right)}{\left(\sqrt{3}+8\right)} = \dfrac{\sqrt{3}+8}{-61}$

19. $\dfrac{2}{\left(3-\sqrt{x}\right)} \cdot \dfrac{\left(3+\sqrt{x}\right)}{\left(3+\sqrt{x}\right)} = \dfrac{2\left(3+\sqrt{x}\right)}{9-x}$

20. $\dfrac{10}{\left(\sqrt{7}-\sqrt{2}\right)} \cdot \dfrac{\left(\sqrt{7}+\sqrt{2}\right)}{\left(\sqrt{7}+\sqrt{2}\right)} = 2\left(\sqrt{7}+\sqrt{2}\right)$

21. $\sqrt{8} - \sqrt{2} = 2\sqrt{2} - \sqrt{2} = \sqrt{2}$

22. $\sqrt{75} + \sqrt{48} = 5\sqrt{3} + 4\sqrt{3} = 9\sqrt{3}$

23. $3\sqrt{72} - \sqrt{98} = 18\sqrt{2} - 7\sqrt{2} = 11\sqrt{2}$

24. $\sqrt{25a} + \sqrt{49a} = 5\sqrt{a} + 7\sqrt{a} = 12\sqrt{a}$

25. $\sqrt{2x^2} - \sqrt{50x^2} = |x|\sqrt{2} - 5|x|\sqrt{2} = -4|x|\sqrt{2}$

26. $\dfrac{10}{\sqrt{5}} + 2\sqrt{45} = 2\sqrt{5} + 6\sqrt{5} = 8\sqrt{5}$

27. $\dfrac{\sqrt{32}}{8} + \dfrac{2}{\sqrt{8}} = \dfrac{4\sqrt{2}}{8} + \dfrac{2\sqrt{8}}{8} = \dfrac{\sqrt{2}}{2} + \dfrac{\sqrt{8}}{4} = \dfrac{\sqrt{2}}{2} + \dfrac{\sqrt{2}}{2} = \sqrt{2}$

28. $x\sqrt{2x} + \sqrt{18x^3} = x\sqrt{2x} + 3x\sqrt{2x} = 4x\sqrt{2x}$

29. $8\sqrt{24} + \dfrac{\sqrt{54}}{3} - 2\sqrt{96} = 16\sqrt{6} + \sqrt{6} - 8\sqrt{6} = 9\sqrt{6}$

30. $6\sqrt{ab^2} + 3b\sqrt{a} - b\sqrt{25a} = 6b\sqrt{a} + 3b\sqrt{a} - 5b\sqrt{a} = 4b\sqrt{a}$

31. $x = 50$

32. $x = 17$

33. $x = 52$

34. $x = 14$

35. $x = 39.4$

36. $x = \dfrac{16}{9}$

37. $x = \dfrac{1}{4}$

38. $6 = x$

39. $x = 13$

40. $x = 2$

Chapter 15

1. $x = \pm 9$

2. $x = \pm 5$

3. $x = \pm 4$

4. $x = 9$ or $x = 5$

5. $x = -6$ or $x = 0$

6. $x = 4$ or $x = 2$

7. $x = 11$

8. $x = \dfrac{11}{3}$ or $x = -1$

9. $x = \dfrac{5}{2}$ $x = -\dfrac{7}{2}$

10. $x = \dfrac{8 \pm 3\sqrt{2}}{5}$

11. $x = 3$ or $x = -7$

12. $t = 2$ or $t = -5$

13. $y = 8$ or $y = -4$

14. $x = 3$ or $x = -2$

15. $w = -3 \pm 3\sqrt{2}$

16. No solution in the real numbers.

17. $x = 1$ or $x = -\dfrac{3}{4}$

18. $x = 1$ or $x = -\dfrac{1}{3}$

19. $x = -3 \pm \sqrt{13}$

20. $x = -2 \pm \sqrt{6}$

21. $b^2 - 4ac = (-8)^2 - 4 \cdot 1 \cdot 12 = 64 - 48 = 16$

22. $b^2 - 4ac = (-3)^2 - 4 \cdot 1 \cdot (-9) = 9 + 36 = 45$

23. $b^2 - 4ac = 5^2 - 4 \cdot 2 \cdot 4 = 25 - 32 = -7$

24. $b^2 - 4ac = 8^2 - 4 \cdot 1 \cdot 16 = 64 - 64 = 0$

25. $b^2 - 4ac = 0^2 - 4 \cdot 1 \cdot (-9) = 36$

26. $b^2 - 4ac = (-5)^2 - 4 \cdot 1 \cdot 4 = 9$ There are two rational solutions.

27. $(-5)^2 - 4 \cdot 2 \cdot (-4) = 57$ There are two irrational solutions.

28. $b^2 - 4ac = 0^2 - 4 \cdot 9 \cdot 4 = -144$ There are no real solutions.

29. $b^2 - 4ac = (-8)^2 - 4 \cdot 1 \cdot 16 = 64 - 64 = 0$ There is one solution.

30. $b^2 - 4ac = (-2)^2 - 4 \cdot 3 \cdot (-1) = 16$ There are two rational solutions.

31. $x = -2$ or $x = -3$

32. $x = 3$ or $x = 4$

33. $x = 4$ or $x = -1$

34. $x = 5$ or $x = 1$

35. $x = -\dfrac{2}{3}$ or $x = 1$

36. $x = -\dfrac{2}{3}$ or $x = \dfrac{3}{2}$

37. $x = -4$ or $x = 2$

38. $x = -5$ or $x = 4$

39. $x = 0$ or $x = -3$

40. $x = \dfrac{1}{2}$ or $x = 1$

Chapter 16

1. $y = x^2 + 4x - 5$
   Vertex: $(-2, -9)$, x-intercepts: $(-5, 0)$ and $(1, 0)$, y-intercept: $(0, -5)$

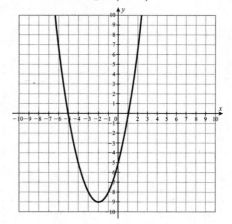

2. $y = x^2 - 2x - 3$
   Vertex: $(1, -4)$, x-intercepts: $(-1, 0)$ and $(3, 0)$, y-intercept: $(0, -3)$

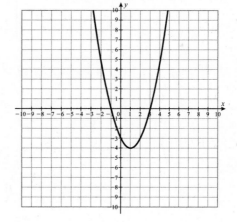

3. $y = 4 - 3x - x^2$

    Vertex: $\left(-\dfrac{3}{2}, \dfrac{25}{4}\right) = (-1.5, 6.25)$,

    x-intercepts: $(-4, 0)$ and $(1, 0)$,

    y-intercept: $(0, 4)$

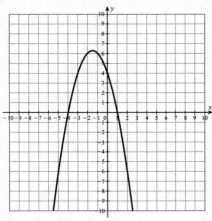

4. $y = x^2 - 8x$

    Vertex: $(4, -16)$,

    x-intercepts: $(0, 0)$ and $(0, 8)$,

    y-intercept: $(0, 0)$

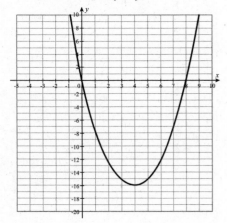

5. $y = 32 + 4x - x^2$

    Vertex: $(2, 36)$, x-intercepts: $(-4, 0)$ and $(8, 0)$,

    y-intercept: $(0, 32)$

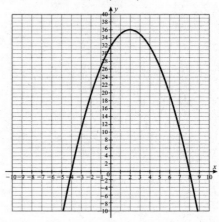

6. $y = x^2 + 4x - 21$

    Vertex: $(-2, -25)$, x-intercepts: $(-7, 0)$ and

    $(3, 0)$, y-intercept: $(0, -21)$

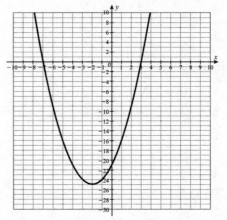

7. $y = 2x^2 - 18$

Vertex: $(0, -18)$, x-intercepts: $(-3, 0)$ and $(3, 0)$, y-intercept: $(0, -18)$

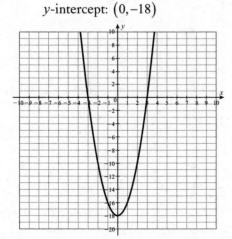

9. $y = -2x^2 - 12x - 18$

Vertex: $(-3, 0)$, x-intercept: $(-3, 0)$, y-intercept: $(0, -18)$

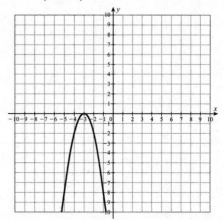

8. $y = 2x^2 + 10x + 12$

Vertex: $\left(-\dfrac{5}{2}, -\dfrac{1}{2}\right)$, x-intercepts: $(-3, 0)$ and $(-2, 0)$, y-intercept: $(0, 12)$

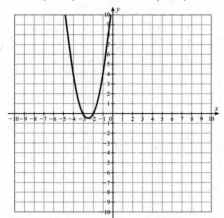

10. $y = 4x^2 + 4x + 1$

Vertex: $\left(-\dfrac{1}{2}, 0\right)$, x-intercept: $\left(-\dfrac{1}{2}, 0\right)$, y=intercept: $(0, 1)$

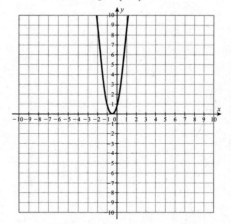

11. $y = 2x^2$

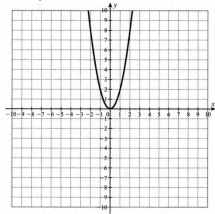

14. $y = x^2 + 6x$

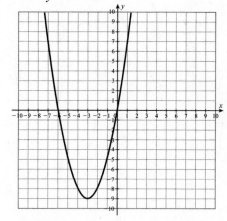

12. $y = 3x^2 - 4$

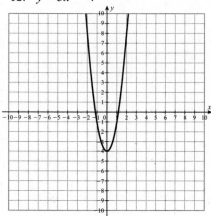

15. $y = x^2 - 2x - 3$

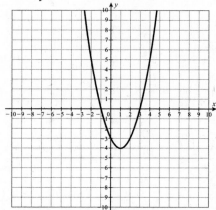

13. $y = 5 - x^2$

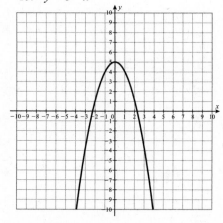

16. $y = 3 - 2x - x^2$

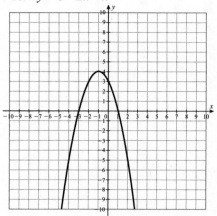

17. $y = 9 - 2x^2$

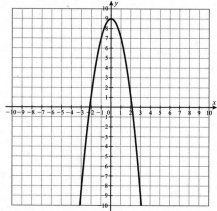

20. $y = 2x^2 - 6x$

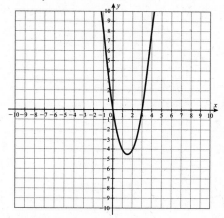

18. $y = x^2 - 1$

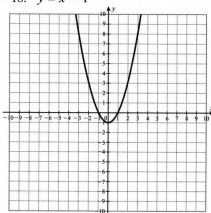

21. $y < -2x^2$

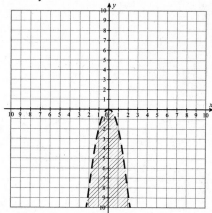

19. $y = 9 - x^2$

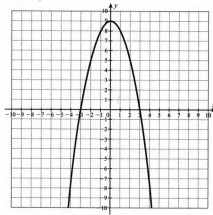

22. $y > x^2 - 6$

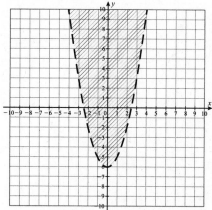

23. $y \leq x^2 + 4x$

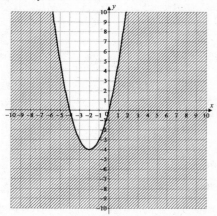

26. $y > x^2 + 4x - 5$

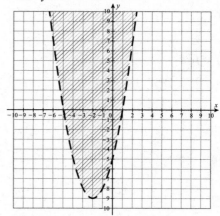

24. $y \geq x^2 - 4$

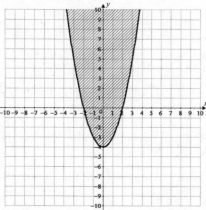

27. $y < 4 - 3x - x^2$

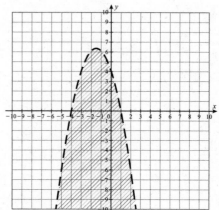

25. $y \leq x^2 - 2x - 3$

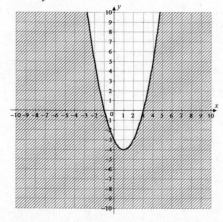

28. $y \geq x^2 + 6x$

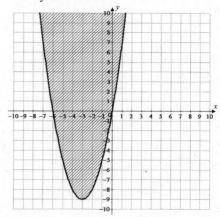

29. $y < x^2 + 4x - 21$

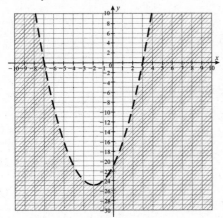

30. $y > 2x^2 + 10x + 12$

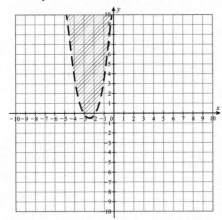

31. $y = 3x^2 - 24x + 45$

32. $y = -2x^2 - 8x - 3$

33. $y = -x^2 + 2$

34. $y = 2x^2 + 16x + 37$

35. $y = 4x^2 + 24x + 34$

36. Vertex $(2,3)$
    $y = 1(x-2)^2 + 3 \Rightarrow y = x^2 - 4x + 7$

37. Vertex $(5,-3)$
    $y = -1(x+5)^2 - 3 \Rightarrow y = -x^2 - 10x - 28$

38. Vertex $(1,-3)$
    $y = 3(x-1)^2 - 3 \Rightarrow y = 3x^2 - 6x$

39. Vertex $(-4,8)$
    $y = -2(x+4)^2 + 8 \Rightarrow y = -2x^2 - 16x - 24$

40. Vertex $(5,-4)$
    $y = \frac{1}{2}(x-5)^2 - 4 \Rightarrow y = \frac{1}{2}x^2 - 5x + \frac{17}{2}$

41. It will take 2.646 seconds to hit the ground.

42. $h(1) = -16 \cdot 1^2 + 112 = 96$ feet high after one second.

43. The object is 76 feet high after 1.5 seconds.

44. After 2 seconds, the object is 261 feet high.

45. The maximum height the object reaches is 405 feet after 5 seconds.

46. There are two moments when the object is 81 feet off the ground, half a second after it is thrown, and again on the way down, 9.5 seconds after it was thrown.

47. The object hits the ground just after 10 seconds.

48. After 0.5 seconds, the object is 66.25 meters from the ground.

49. The object is 20 meters high at approximately 2.14 seconds after it was released.

50. It takes approximately 2.65 seconds to hit the ground.

Chapter 17

1. All real numbers except 0.

2. All real numbers except 7.

3. All real numbers except -8.

4. All real numbers except 3.

5. All real numbers except 1 and -1.

6. All real numbers except -2.

7. All real numbers except -3.

8. All real numbers except 0 and -5.

9. All real numbers except -2 and -3.

10. All real numbers. (There is no value of x that can make the denominator equal 0.)

11. $\dfrac{9\cancel{x^2}}{48\cancel{x^2}} = \dfrac{\overset{3}{\cancel{9}}}{\underset{16}{\cancel{48}}} = \dfrac{3}{16}$, $x \neq 0$

12. $\dfrac{52x}{13x^2} = \dfrac{4}{x}$, $x \neq 0$

13. $\dfrac{3x+6}{x+2} = \dfrac{3\cancel{(x+2)}}{\cancel{x+2}} = 3$, $x \neq -2$

14. $\dfrac{3x+6}{15x-12} = \dfrac{\cancel{3}(x+2)}{\cancel{3}(5x-4)} = \dfrac{x+2}{5x-4}$, $x \neq \dfrac{4}{5}$

15. $\dfrac{x^2-6x}{x^2-7x+6} = \dfrac{x\cancel{(x-6)}}{\cancel{(x-6)}(x-1)} = \dfrac{x}{x-1}$, $x \neq 1,6$

16. $\dfrac{x^2-x-42}{x^2-49} = \dfrac{\cancel{(x-7)}(x+6)}{\cancel{(x-7)}(x+7)} = \dfrac{x+6}{x+7}$, $x \neq \pm 7$

17. $\dfrac{x^2-8x+15}{x^2-7x+10} = \dfrac{(x-3)\cancel{(x-5)}}{\cancel{(x-5)}(x-2)} = \dfrac{x-3}{x-2}$, $x \neq 2,5$

18. $\dfrac{2x^2+5x+2}{(2x+1)^2} = \dfrac{\cancel{(2x+1)}(x+2)}{\cancel{(2x+1)}(2x+1)} = \dfrac{x+2}{2x+1}$, $x \neq -\dfrac{1}{2}$

19. $\dfrac{x^2-4x}{x^2-16} = \dfrac{x\cancel{(x-4)}}{(x+4)\cancel{(x-4)}} = \dfrac{x}{x+4}$, $x \neq \pm 4$

20. $\dfrac{3x^2+6x+3}{6x^2+18x+12} = \dfrac{\overset{1}{\cancel{3}}\cancel{(x+1)}(x+1)}{\underset{2}{\cancel{6}}\cancel{(x+1)}(x+2)} = \dfrac{x+1}{2(x+2)} = \dfrac{x+1}{2x+4}$, $x \neq -1,-2$

21. $\dfrac{3(x+5)^2}{\cancel{15}} \cdot \dfrac{\cancel{5}}{x+5} = \dfrac{(x+5)\cancel{(x+5)}}{\cancel{(x+5)}} = x+5$

22. $\dfrac{x^2-16}{2} \cdot \dfrac{8}{x+4} = \dfrac{\cancel{(x+4)}(x-4)}{\cancel{2}} \cdot \dfrac{\overset{4}{\cancel{8}}}{\cancel{x+4}} = 4(x-4) = 4x-16$

23. $\dfrac{x^2+2x+1}{5x-10} \cdot \dfrac{2x-4}{x^2-1} = \dfrac{\cancel{(x+1)}(x+1)}{5\cancel{(x-2)}} \cdot \dfrac{2\cancel{(x-2)}}{\cancel{(x+1)}(x-1)} = \dfrac{2(x+1)}{5(x-1)} = \dfrac{2x+2}{5x-5}$

24. $\dfrac{x^2+7x+10}{x^2-2x+1} \cdot \dfrac{x^2-6x+5}{x^2-25} = \dfrac{(x+2)\cancel{(x+5)}}{(x-1)\cancel{(x-1)}} \cdot \dfrac{\cancel{(x-1)}\cancel{(x-5)}}{\cancel{(x+5)}\cancel{(x-5)}} = \dfrac{x+2}{x-1}$

25. $\dfrac{x^2-5x+6}{x^2-7x+10} \cdot \dfrac{x^2-4x-5}{x^2-2x-3} = \dfrac{\cancel{(x-2)}\cancel{(x-3)}}{\cancel{(x-2)}\cancel{(x-5)}} \cdot \dfrac{\cancel{(x-5)}\cancel{(x+1)}}{\cancel{(x+1)}\cancel{(x-3)}} = 1$

26. $\dfrac{5x^2-20}{x^2-5x+6} \div \dfrac{6x+12}{3x-5} = \dfrac{5\cancel{(x+2)}(x-2)}{\cancel{(x-2)}(x-3)} \cdot \dfrac{3x-5}{6\cancel{(x+2)}} = \dfrac{5(3x-5)}{6(x-3)} = \dfrac{15x-25}{6x-18}$

27. $\dfrac{x^2-4}{2x+5} \div \dfrac{2x-4}{4x^2-25} = \dfrac{x^2-4}{2x+5} \cdot \dfrac{4x^2-25}{2x-4} = \dfrac{(x+2)(2x-5)}{2} = \dfrac{2x^2-x-10}{2}$

28. $\dfrac{x^2+x-42}{x^2-36} \div \dfrac{x^2-x-42}{x^2-49} = \dfrac{x^2+x-42}{x^2-36} \cdot \dfrac{x^2-49}{x^2-x-42} = \dfrac{(x+7)^2}{(x+6)^2} = \dfrac{x^2+14x+49}{x^2+12x+36}$

29. $\dfrac{x^2-1}{3x-6} \div \dfrac{6x+6}{8x-16} = \dfrac{\cancel{(x+1)}(x-1)}{3\cancel{(x-2)}} \cdot \dfrac{\overset{4}{\cancel{8}}\cancel{(x-2)}}{\underset{3}{\cancel{6}}\cancel{(x+1)}} = \dfrac{4(x-1)}{9} = \dfrac{4x-4}{9}$

30. $\dfrac{2x^2+5x+2}{6x+3} \div \dfrac{2x^2+x-6}{2x-3} = \dfrac{\cancel{(2x+1)}\cancel{(x+2)}}{3\cancel{(2x+1)}} \cdot \dfrac{\cancel{2x-3}}{\cancel{(2x-3)}\cancel{(x+2)}} = \dfrac{1}{3}$

31. $\dfrac{t}{t+5} + \dfrac{5}{t+5} = \dfrac{t+5}{t+5} = 1$

32. $\dfrac{5x+1}{x+3} + \dfrac{5-3x}{x+3} = \dfrac{2x+6}{x+3} = \dfrac{2\cancel{(x+3)}}{\cancel{x+3}} = 2$

33. $\dfrac{5x}{x^2-4} - \dfrac{10}{x^2-4} = \dfrac{5x-10}{x^2-4} = \dfrac{5\cancel{(x-2)}}{(x+2)\cancel{(x-2)}} = \dfrac{5}{x+2}$

34. $\dfrac{3x+2}{7x} + \dfrac{5x+7}{7x} = \dfrac{8x+9}{7x}$

35. $\dfrac{2y-3}{9y} - \dfrac{5y-4}{9y} = \dfrac{-3y+1}{9y}$

36. $\dfrac{12x-5}{2x+5} - \dfrac{8x-5}{2x+5} = \dfrac{4x}{2x+5}$

37. $\dfrac{x^2-7}{x-y} - \dfrac{y^2-7}{x-y} = \dfrac{x^2-y^2}{x-y} = \dfrac{(x+y)\cancel{(x-y)}}{\cancel{x-y}} = x+y$

38. $\dfrac{2x^2+9x+4}{x^2-5x+6} - \dfrac{x^2+15x-5}{x^2-5x+6} = \dfrac{x^2-6x+9}{x^2-5x+6} = \dfrac{(x-3)\cancel{(x-3)}}{(x-2)\cancel{(x-3)}} = \dfrac{x-3}{x-2}$

39. $\dfrac{3x-2}{5x}+\dfrac{2x-3}{5x}-\dfrac{5-2x}{5x}=\dfrac{5x-5}{5x}-\dfrac{5-2x}{5x}=\dfrac{7x-10}{5x}$

40. $\dfrac{2x+19}{x^2+3x+2}+\dfrac{7x-16}{x^2+3x+2}-\dfrac{3x-9}{x^2+3x+2}=\dfrac{6x+12}{x^2+3x+2}=\dfrac{6\cancel{(x+2)}}{(x+1)\cancel{(x+2)}}=\dfrac{6}{x+1}$

41. $\dfrac{a}{a+3}+\dfrac{2a}{3(a+3)}=\dfrac{3a}{3(a+3)}+\dfrac{2a}{3(a+3)}=+\dfrac{5a}{3(a+3)}$

42. $\dfrac{7}{3x+6}-\dfrac{2}{5x+10}=\dfrac{35}{15(x+2)}-\dfrac{6}{15(x+2)}=\dfrac{29}{15(x+2)}$

43. $\dfrac{5}{7}-\dfrac{2}{x+7}=\dfrac{5(x+7)}{7(x+7)}-\dfrac{2\cdot7}{7(x+7)}=\dfrac{5x+35-14}{7(x+7)}=\dfrac{5x+21}{7(x+7)}$

44. $\dfrac{x}{x+5}+\dfrac{5}{x-5}=\dfrac{x(x-5)}{(x+5)(x-5)}+\dfrac{5(x+5)}{(x+5)(x-5)}=\dfrac{x^2-5x+5x+25}{(x+5)(x-5)}=\dfrac{x^2+25}{(x+5)(x-5)}$

45. $\dfrac{8}{x+6}+\dfrac{7}{x^2+6x}=\dfrac{8x}{x(x+6)}+\dfrac{7}{x(x+6)}=\dfrac{8x+7}{x(x+6)}$

46. $\dfrac{x+2}{x+3}+\dfrac{x+3}{x+4}=\dfrac{(x+2)(x+4)}{(x+3)(x+4)}+\dfrac{(x+3)(x+3)}{(x+4)(x+3)}=\dfrac{2x^2+12x+17}{(x+4)(x+3)}$

47. $\dfrac{y-7}{y-9}-\dfrac{y-2}{y-3}=\dfrac{(y-7)(y-3)}{(y-9)(y-3)}-\dfrac{(y-2)(y-9)}{(y-3)(y-9)}=\dfrac{y+3}{(y-3)(y-9)}$

48. $\dfrac{6x}{x^2+5x+6}-\dfrac{2}{x+3}=\dfrac{6x}{(x+3)(x+2)}-\dfrac{2(x+2)}{(x+3)(x+2)}=\dfrac{4x-4}{(x+3)(x+2)}$

49. $\dfrac{4}{x^2+7x+10}-\dfrac{1}{x^2-25}=\dfrac{3x-22}{(x+2)(x+5)(x-5)}$

50. $\dfrac{2a+3}{a^2+4a+4}-\dfrac{a-2}{a^2+5a+6}=\dfrac{2a^2+9a+9-(a^2-4)}{(a+2)(a+2)(a+3)}=\dfrac{a^2+9a+13}{(a+2)(a+2)(a+3)}$

Chapter 18

1. $\dfrac{8}{3x}=2\Rightarrow8=6x\Rightarrow x=\dfrac{8}{6}=\dfrac{4}{3}$

5. $\dfrac{7x-4}{6x+4}=\dfrac{6}{7}\Rightarrow49x-28=36x+24\Rightarrow13x=52\Rightarrow x=4$

2. $\dfrac{3-5x}{6}=3\Rightarrow3-5x=18\Rightarrow-5x=15\Rightarrow x=-3$

6. $\dfrac{3}{5x-1}=\dfrac{2}{3x}\Rightarrow9x=10x-2\Rightarrow-x=-2\Rightarrow x=2$

3. $\dfrac{4x-6}{x}=2\Rightarrow4x-6=2x\Rightarrow-6=-2x\Rightarrow x=3$

4. $\dfrac{a}{a-5}=\dfrac{8}{3}\Rightarrow8a-40=3a\Rightarrow5a=40\Rightarrow a=8$

7. $\dfrac{1}{x} = \dfrac{1}{x^2 - 2x} \Rightarrow x^2 - 2x = x \Rightarrow x^2 - 3x = 0 \Rightarrow x(x-3) = 0 \Rightarrow x = 0, x = 3$

   The solution $x = 0$ must be rejected, so the solution is $x = 3$.

8. $\dfrac{20}{3x^2 - 6x} = \dfrac{5}{x} \Rightarrow 15x^2 - 30x = 20x \Rightarrow 15x^2 - 50x = 0 \Rightarrow 5x(3x - 10) = 0 \Rightarrow x = 0, x = \dfrac{10}{3}$

   The solution $x = 0$ must be rejected, so the solution is $x = \dfrac{10}{3}$.

9. $\dfrac{5}{2x} = \dfrac{7}{2x - 8} \Rightarrow 14x = 10x - 40 \Rightarrow 4x = -40 \Rightarrow x = -10$

10. $\dfrac{-1}{x-2} = \dfrac{2}{x^2 - 4} \Rightarrow -x^2 + 4 = 2x - 4 \Rightarrow 0 = x^2 + 2x - 8 \Rightarrow (x+4)(x-2) = 0 \Rightarrow x = -4, x = 2$

    The solution $x = 2$ must be rejected, so the solution is $x = -4$.

11. $3x\left(\dfrac{1}{x} + \dfrac{1}{3x}\right) = 28 \cdot 3x \Rightarrow 3 + 1 = 84x \Rightarrow 4 = 84x \Rightarrow x = \dfrac{4}{84} = \dfrac{1}{21}$

12. $4x\left(\dfrac{5}{x} + \dfrac{15}{2x}\right) = \left(\dfrac{5}{4}\right) \cdot 4x \Rightarrow 20 + 30 = 5x \Rightarrow 5x = 50 \Rightarrow x = 10$

13. $x = 10$

14. $x = \dfrac{5}{3}$

15. $x = 3$

16. $t = 5$

17. $y = 8$

18. $1 = x$

    The solution of $x = 1$ must be rejected, so there is no solution.

19. $x = -9$

20. $x = 4$

21. $y = \dfrac{4}{x}$ Asymptotes: $x = 0, \quad y = 0,$
    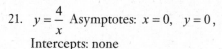
    Intercepts: none

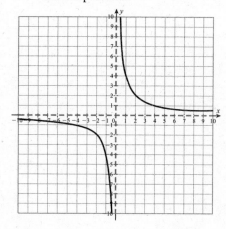

22. $y = \dfrac{-12}{x}$ Asymptotes: $x = 0$, $y = 0$,
    Intercepts: none

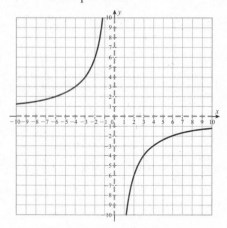

24. $y = \dfrac{x}{x+4}$ Asymptotes: $x = -4$, $y = 1$,
    Intercepts: $(0,0)$

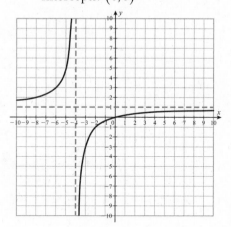

23. $y = \dfrac{3}{x-2}$ Asymptotes: $x = 2$, $y = 0$,
    Intercepts: $\left(0, -\dfrac{3}{2}\right)$

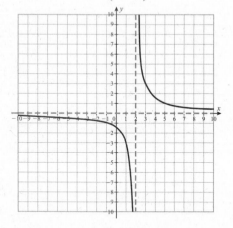

25. $y = \dfrac{-6}{x-3}$ Asymptotes: $x = 3$, $y = 0$,
    Intercepts: $(0,2)$

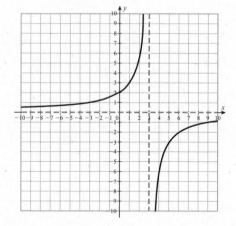

26. $y = \dfrac{x-1}{x+5}$ Asymptotes: $x = -5$, $y = 1$,

   Intercepts: $\left(0, -\dfrac{1}{5}\right)$, $(1, 0)$

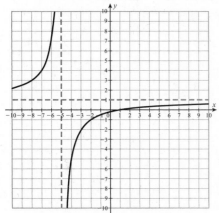

28. $y = \dfrac{2x+6}{x-1}$ Asymptotes: $x = 1$, $y = 2$,

   Intercepts: $(0, -6)$, $(-3, 0)$

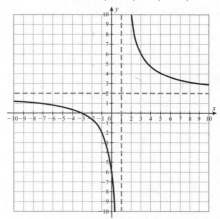

27. $y = \dfrac{x+3}{x-4}$ Asymptotes: $x = 4$, $y = 1$,

   Intercepts: $\left(0, -\dfrac{3}{4}\right)$, $(-3, 0)$

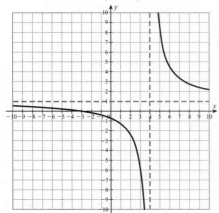

29. $y = \dfrac{x+5}{2x}$ Asymptotes: $x = 0$, $y = \dfrac{1}{2}$,

   Intercepts: $(-5, 0)$

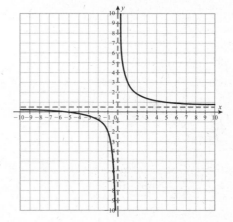

30. $y = \dfrac{2x}{x+5}$ Asymptotes: $x = -5$, $y = 2$,

Intercepts: $(0,0)$

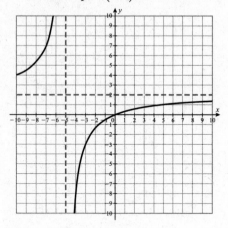

Index

F

G

H

I-J-K

L

U–V

W–X–Y–Z

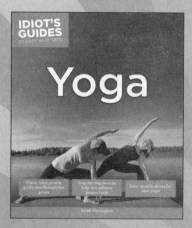

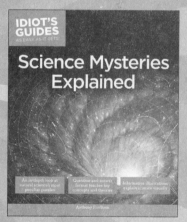
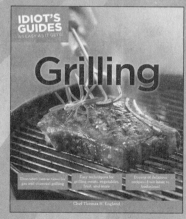

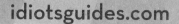

There's a lot of crummy "how-to" content out there on the internet. A LOT. We want to fix that, and YOU can help!

31901064503156